AF595127

# The Application of Mathematics to Physics and Nonlinear Science

# The Application of Mathematics to Physics and Nonlinear Science

Special Issue Editor

**Andrei Ludu**

MDPI • Basel • Beijing • Wuhan • Barcelona • Belgrade • Manchester • Tokyo • Cluj • Tianjin

*Special Issue Editor*
Andrei Ludu
Embry-Riddle Aeronautical University
USA

*Editorial Office*
MDPI
St. Alban-Anlage 66
4052 Basel, Switzerland

This is a reprint of articles from the Special Issue published online in the open access journal *Mathematics* (ISSN 2227-7390) (available at: https://www.mdpi.com/journal/mathematics/special_issues/physics_nonlinear_science).

For citation purposes, cite each article independently as indicated on the article page online and as indicated below:

LastName, A.A.; LastName, B.B.; LastName, C.C. Article Title. *Journal Name* **Year**, *Article Number*, Page Range.

**ISBN 978-3-03928-726-0 (Pbk)**
**ISBN 978-3-03928-727-7 (PDF)**

# Contents

# About the Special Issue Editor

**Andrei Ludu** (Professor of Mathematics, Ph.D.) graduated with a M.S. in Theoretical Physics at Bucharest University in Romania in 1980 and defended his Ph.D. in 1988 at the Institute of Atomic Physics, Bucharest-Magurele, with a thesis on Lie groups methods in hot and dense thermonuclear plasmas. During 1981–1987, he worked as senior researcher at the National Thermonuclear Program H in Romania. He was Associate Professor at the Theoretical Physics and Mathematics Dept. at Bucharest University until 1992, when he moved as Guest Professor at Goethe-University and J. Liebig University in Frankfurt/Main and Giessen, respectively. He was several times an invited scientist at Los Alamos National Laboratory; ICTP Trieste, Italy; Turku University, Finland; Kurcheatov Institute, Moscow; and at Antwerp and ULB Universities in Belgium. In 1996, he moved to the USA and occupied positions of Senior Researcher and Professor of Physics at Louisiana State University's Physics and Astronomy Dept., and Northwestern State University's Physics and Chemistry Dept., respectively. Since 2011, he has been Professor of Math and Director of the Wave Lab at the Math Dept. at Embry-Riddle Aeronautical University in Daytona Beach, Florida, where he works together with his wife Maria, also a Math Professor. Dr. Ludu has published 70 articles in peer-reviewed journals, 2 monograph books with Springer, 2 more books in collaboration, and has more than 200 published contributions at international science conferences. He has 2 patents of invention, is founder of the IDEAS Program at Northwestern State University, designed and built nonlinear wave labs at three universities, and worked as research consultant with Procter & Gamble Co., Hydro-Plus Engineering, Cardinal Systems, and GFS Corp. He proved the existence of magnetic field solitons in thermonuclear plasma; explained in 1993, together with Greiner and Sandulescu, the emission of alpha particles from heavy nuclei as nuclear surface solitons; proved that the algebraic structure of Haar and Daubechies wavelets is a quantum deformation of the Fourier system generating algebra; discovered rotons (as solitons moving around circles), which were experimentally proven in 2019 to exist; discovered new modes of swimming of T. Brucei as solitary depression waves; has substantial contributions in the discovery of spontaneously rotating Leidenfrost hollow polygons; and recently has introduced the time variable order of differentiation equations.

# Preface to "The Application of Mathematics to Physics and Nonlinear Science"

The importance of understanding nonlinearity has increased over the decades through the development of newer fields of application: biophysics, wave dynamics, optical fibers, plasmas, ecological systems, micro fluids, and cross-disciplinary fields. The necessary mathematics involves nonlinear evolution equations. Obtaining closed-form solutions for these equations plays an important role in the proper understanding of features of many phenomena by unraveling their complex mechanisms such as pattern formation and selection, the spatial localization of transfer processes, the multiplicity or absence steady states under various conditions, the existence of peaking regimes, etc. Even exact test solutions with no immediate physical meaning are used to verify the consistency of numerical, asymptotic, and approximate analytical methods. This Special Issue gathered a few of the most important topics in nonlinear science.

The problem of nonlinear viscoelastic fluid flow has been studied extensively by different mathematicians over the past several decades starting in the 1970s. The mathematical model describing such flows, including liquid polymers dynamics, is given by the Navier–Stokes–Voigt (also called Kelvin–Voigt, or weakly compressible) equations. Although their local-in-time solvability and weak solutions existence and uniqueness in the framework of the Hilbert space techniques were established for several special configurations (blowup of solutions, various slip problems, weak solutions of the g-Kelvin–Voigt equations for viscoelastic fluid flows in thin domains, Dirichlet problems, inverse problem, coupled system of nonlinear equations for heat transfer in steady-state flows of a polymeric fluid, etc.), the relevant question for the existence and uniqueness of strong solutions in a Banach space, under natural conditions, was not previously solved. E. S. Baranovskii, in this Special Issue, proves the existence and uniqueness of a strong solution to the incompressible Navier–Stokes–Voigt model as a nonlinear evolutionary equation in suitable Banach space. In addition, convenient algorithms for finding these strong solutions in 2D and 3D domains are developed by using the Faedo–Galerkin procedure with a special basis of eigenfunctions of the Stokes operator and deriving various a priori estimates of approximate solutions in Sobolev's spaces.

The nonlinear behavior of complex fluids and soft matter (interfacial fluid flow, polymer science, and in industrial applications) are important topics in the present front of the wave research. The Cahn–Hilliard equations model some of these phenomena, especially when the system consists of binary mixtures or more generally for interface-related problems, such as the spinodal decomposition of a binary alloy mixture, in painting of binary images, microphase separation of co-polymers, microstructures with elastic inhomogeneity, two-phase binary fluids, in silico tumor growth simulation, and structural topology optimization. Some of these problem are related to solutions of Stefan problems and the model of Thomas and Windle for diffuse interface problems. Another interesting application is for the coupling of the phase separation of the Cahn–Hilliard equation to the Navier–Stokes equations of fluid flow. In this phase separation, the two components of a binary fluid spontaneously separate and form domains pure in each component. C. Lee et al. studied the 2D Cahn–Hilliard equation numerically using a novel nonlinear multigrid method.

Another important topic in this Special Issue is represented by the article authored by J. Danane et al. on a mathematical model describing viral dynamics in the presence of the latently infected cells and the cytotoxic T-lymphocytes cells, considering the spatial mobility of free viruses. Mathematical modelling becomes an important tool for the understanding and predicting the spread of viral infection and for the development of efficient strategies to control its dynamics. Viral infections represent a major cause of morbidity with important consequences for patient health and society. Among the most dangerous are the human immunodeficiency virus that attacks immune cells leading to the deficiency of the immune system, the human papillomavirus that infects basal cells of the cervix, and the hepatitis B/C viruses that attack liver cells. In this paper, the authors couple five nonlinear differential equations describing the interaction among the uninfected cells, the latently infected cells, the actively infected cells, the free viruses, and the cellular immune response. The existence, positivity, boundedness, and global stability of each steady state obtained through Lyapunov functionals for the suggested diffusion model are proved. The theoretical results are validated by numerical simulations for each case.

In the paper by D. Dutych, the Feller's nonlinear diffusion equation and its numerical solutions are analyzed. This equation arises naturally in probability and physics (e.g., wave turbulence theory). In previous literature, this equation was discretized naively. This approach may introduce serious numerical difficulties since the diffusion coefficient is practically unbounded and most of its solutions are weakly divergent at the origin. To overcome these difficulties, the author reformulated this equation using inspiration from Lagrangian fluid mechanics.

Another interesting topic included in this Special Issue concerns nonlinear evolution equations modeling waves in rotating fluids. The case of finite depth fluid and small-amplitude long waves is analyzed in the frame of the Ostrovsky model. This model generalizes the Korteweg–deVries equation by the additional term induced by the Coriolis force. There are several studies in the literature on the local and global well-posedness in energy space, stability of solitary waves, and convergence of solutions in the limit of the Korteweg–deVries equation. In the paper authored by G. M. Coclite and L. di Ruvo, the reduced Ostrovsky equation (also known as the Ostrovsky–Hunter equation, the short-wave equation, or the Vakhnenko equation) is analyzed. The authors discuss the continuum spectrum pulse equation as a third-order nonlocal nonlinear evolution equation related to the dynamics of the electrical field of linearly polarized continuum spectrum pulses in optical waveguides. The authors study the well-posedness of the classical solutions to the Cauchy problem associated with this equation.

**Andrei Ludu**
*Special Issue Editor*

*Article*

# Numerical Simulation of Feller's Diffusion Equation

**Denys Dutykh** [1,2]

[1] Univ. Grenoble Alpes, Univ. Savoie Mont Blanc, CNRS, LAMA, 73000 Chambéry, France; Denys.Dutykh@univ-smb.fr or Denys.Dutykh@gmail.com; Tel.: +33-04-79-75-94-38

[2] LAMA, UMR 5127 CNRS, Université Savoie Mont Blanc, Campus Scientifique, F-73376 Le Bourget-du-Lac CEDEX, France

Received: 12 September 2019; Accepted: 4 November 2019; Published: 6 November 2019

**Abstract:** This article is devoted to FELLER's diffusion equation, which arises naturally in probability and physics (e.g., wave turbulence theory). If discretized naively, this equation may represent serious numerical difficulties since the diffusion coefficient is practically unbounded and most of its solutions are weakly divergent at the origin. In order to overcome these difficulties, we reformulate this equation using some ideas from the LAGRANGIAN fluid mechanics. This allows us to obtain a numerical scheme with a rather generous stability condition. Finally, the algorithm admits an elegant implementation, and the corresponding MATLAB code is provided with this article under an open source license.

**Keywords:** Feller equation; parabolic equations; Lagrangian scheme; Fokker–Planck equation; probability distribution

**PACS:** 02.30.Jr (primary); 02.60.Cb; 02.50.Cw (secondary)

**MSC:** 35K20 (primary); 65M06; 65M75 (secondary)

---

## 1. Introduction

The celebrated FELLER equation was introduced in two seminal papers published by William FELLER (1951/1952) in Annals of Mathematics [1,2]. These publications studied mathematically (and, henceforth, gave the name) the following equation (To be more accurate, W. FELLER studied the following equation [1]:

$$p_t = \left[ a\,x\,u \right]_{xx} - \left[ (c + b\,x)\,u \right]_x,$$

where $a > 0$ and $0 < x < +\infty$.).

$$p_t + \mathfrak{F}_x = 0, \qquad \mathfrak{F}(p, x, t) \stackrel{\text{def}}{:=} -x \cdot (\gamma\, p + \eta\, p_x), \tag{1}$$

where the subscripts $t$, $x$ denote the partial derivatives, i.e., $(\cdot)_t \stackrel{\text{def}}{:=} \dfrac{\partial(\cdot)}{\partial t}$, $(\cdot)_x \stackrel{\text{def}}{:=} \dfrac{\partial(\cdot)}{\partial x}$. Two parameters $\gamma$ and $\eta > 0$ can be time-dependent in some physical applications, even if in this study we assume they are constants, for the sake of simplicity (The numerical method we are going to propose can be straightforwardly generalized for this case when $\gamma = \gamma(t)$ and $\eta = \eta(t)$. Moreover, FELLER's processes with time-varying coefficients were studied recently in [3].). Equation (1) can be seen as the FOKKER–PLANCK (or the forward KOLMOGOROV) equation, with $\gamma\,x$ being the drift and $\eta\,x$ being the diffusion coefficients (see [4] for more information on the FOKKER–PLANCK equation). One can notice also that Equation (1) becomes singular at $x = 0$ and $x = +\infty$. We remind that practically important solutions to FELLER's equation might be unbounded near $x = 0$. In order to attempt solving Equation (1), one has to prescribe an initial condition $p(x, 0) = p_0(x)$, presumably with a boundary condition at $x = 0$. A popular choice is to prescribe the homogeneous boundary condition $p(0, t) \equiv 0$. For this choice of the boundary condition, it is not difficult to show that the FELLER equation dynamics would preserve solution positivity provided that $p_0(x) \geqslant 0$ (see Appendix A for a proof). The solution norm is also preserved (see Appendix B). Moreover, using the LAPLACE transform techniques, FELLER has shown in [1] that the initial condition $p_0(x)$ determines uniquely the solution. In other words, *no* boundary condition at $x = 0$ should be prescribed. This conclusion might appear, perhaps, to be counter-intuitive.

The great interest in FELLER's equation can be explained by its connection to FELLER's processes, which can be described by the following stochastic differential LANGEVIN equation (The stochastic differential equations are understood in the sense of ITŌ.):

$$\mathrm{d}X_t = -\gamma X_t \mathrm{d}t + \sqrt{2\eta X_t}\,\mathrm{d}\mathcal{W}_t,$$

where $\mathcal{W}_t$ is the standard WIENER process, i.e., $\xi(t) \stackrel{\text{def}}{:=} \dfrac{\mathrm{d}\mathcal{W}_t}{\mathrm{d}t}$ is zero-mean GAUSSIAN white noise, i.e.,

$$\langle \xi(t) \rangle = 0, \qquad \langle \xi(t)\, \xi(s) \rangle = \delta(t - s),$$

where the brackets $\langle \cdot \rangle$ denote an ensemble averaging operator. Then, the Probability Density Function (PDF) $p(x, t; x_0)$ of the process $X(t)$, i.e.,

$$\mathbb{P}\left\{x < X(t) < x + \mathrm{d}x \mid X(0) = x_0\right\} \equiv p(x, t; x_0)\,\mathrm{d}x,$$

satisfies Equation (1) with the following initial condition [5]:

$$p_0(x) = \delta(x - x_0), \qquad x_0 \in \mathbb{R}^+.$$

The point $x = 0$ is a singular boundary that the process $X(t)$ cannot cross. The FELLER process is a continuous representation of branching and birth–death processes, which never attains negative values. This property makes it an ideal model not only in physical, but also in biological and social sciences [3,6,7].

As a general comprehensive reference on generalized FELLER's equations, we can mention the book [8]. Since at least a couple of years ago there has again been a growing interest for studying Equation (1). Some singular solutions to FELLER's equation with constant coefficients were constructed in [6] via spectral decompositions. FELLER's equation and FELLER's processes with time-varying coefficients were studied analytically (always using the LAPLACE transform) and asymptotically in [3].

The FELLER (and FOKKER–PLANCK) equation 'has already made its appearance in optical communications [9]. Recently, the FELLER equation was derived in the context of the weakly interacting random waves dominated by four-wave interactions [10]. Wave Turbulence (We could define the Wave Turbulence (WT) as a physical and mathematical study of systems where random and coherent waves coexist and interact [11].) WT is a common name for such processes [11]. In WT, the FELLER equation governs the PDF of squared FOURIER wave amplitudes, i.e., $x \sim |a|^2$. In [10], some steady solutions to this equation with finite flux in the amplitude space were constructed (There is probably a misprint in ([10] p. 366). To obtain mathematically correct solutions, one has to define $n_k \stackrel{\text{def}}{:=} \dfrac{\eta}{\gamma}$ on the line below Equation (14)). See also [12], Chapter 11 for a detailed discussion and interpretations. Recently, the FELLER equation has been studied analytically in [13]. The authors applied the LAPLACE transform to it in space (this computation can be found even earlier in [1], Equation (3.1)) and the resulting non-homogeneous hyperbolic equation was solved using the method of characteristics along the lines presented in ([1], Section §3) (see ([1], Equation (3.9)) for the general analytical solution).

The behaviour of solutions $p(x, t)$ for large $x$ describes the appearance probability of extreme waves. In the context of ocean waves, these extreme events are known as *rogue* (or *freak*) waves [14]. In the WT literature, any noticeable deviation from the RAYLEIGH distribution for $x \gg 1$ is referred to as the *anomalous probability distribution* of large amplitude waves [10]. For GAUSSIAN wave fields, all statistical properties can be derived from the spectrum. However, the PDFs and other higher order moments are compulsory tools to study such deviations.

The present study focuses on the numerical discretization and simulation of FELLER equation. The naive approach to solve this equation numerically encounters notorious difficulties. The first question that arises is what is the (numerical) boundary condition to be imposed at $x = 0$? Moreover, one can notice that Equation (1) is posed on a semi-infinite domain. There are three main strategies to tackle this difficulty:

1. Map $\mathbb{R}^+$ on a finite interval $[0, \ell]$;
2. Use spectral expansions on $\mathbb{R}^+$ (e.g., LAGUERRE or associated LAGUERRE polynomials);
3. Replace (truncate) $\mathbb{R}^+$ to $[0, L]$, with $L \gg 1$.

In most studies, the latter option is retained by imposing some appropriate boundary conditions at the artificial boundary $x = L$. In our study, we shall propose a method that is able to handle the semi-infinite domain $\mathbb{R}^+$ without any truncations or simplifications. Finally, the diffusion coefficient in the FELLER Equation (1) is unbounded. If the domain is truncated at $x = L$, then the diffusion coefficient takes the maximal value $\nu_{\max} \stackrel{\text{def}}{:=} \eta L \gg 1$, which depends on the truncation limit $L$ and can become very large in practice. We remind also that explicit schemes for diffusion equations are subject to the so-called COURANT–FRIEDRICHS–LEWY (CFL) stability conditions [15]:

$$\Delta t \leqslant \frac{\Delta x^2}{2\,\nu_{\max}}.$$

Taking into account the fact that $\nu_{\max}$ can be arbitrarily large, no explicit scheme can be usable with FELLER equation in practice. Moreover, the dynamics of the FELLER equation spread over the space $\mathbb{R}^+$ even localized initial conditions. In general, one can show that the support of $p(x, t)$, $t > 0$ is strictly larger (Using modern analytical techniques, it is possible to show even sharper results on the solution support, see e.g., [16].) than the one of $p(x, 0)$. It is the so-called *retention property*. Thus, longer simulation times require larger domains. For all these reasons, it becomes clear that numerical discretization of the FELLER equation requires special care.

In this study, we demonstrate how to overcome this assertion as well. The main idea behind our study is to bring together PDEs and Fluid Mechanics. First, we observe that the classical EULERIAN description is not suitable for this equation, even if the problem is initially formulated in the EULERIAN setting. Consequently, the FELLER equation will be recast in special *material* or the so-called LAGRANGIAN variables (It is known that both EULERIAN and LAGRANGIAN descriptions were proposed by the same person, Leonhard EULER), which make the resolution easier and naturally adaptive ([17], Chapter 7).

The present manuscript is organized as follows. The symmetry analysis of Equation (1) is performed in Section 2. Then, the governing equation is reformulated in LAGRANGIAN variables in Section 3. The numerical results are presented in Section 4. Finally, the main conclusions and perspectives are outlined in Section 5.

## 2. Symmetry Analysis

In general, a linear PDE admits an infinity of conservation laws, with integrating multipliers being solutions to the adjoint PDE [18]. Here, we provide an interesting conservation law, which was found using the GEM MAPLE package [19]:

$$\Bigl(\mathrm{E}_1\,(-\frac{\gamma\,x}{\eta})\,p\Bigr)_t + \mathfrak{G}_x = 0,$$

where $\mathrm{E}_1(z) \stackrel{\text{def}}{:=} \int_1^{+\infty} \frac{\mathrm{e}^{-tz}}{t}\,\mathrm{d}t$ is the so-called *exponential integral* function [20] and the flux $\mathfrak{G}$ is defined as

$$\mathfrak{G}(x, p) \stackrel{\text{def}}{:=} -\eta\,\mathrm{e}^{\frac{\gamma x}{\eta}}\,p - x\Bigl(\gamma\,\mathrm{E}_1\bigl(-\frac{\gamma x}{\eta}\bigr)\,p + \eta\,\mathrm{E}_1\bigl(-\frac{\gamma x}{\eta}\bigr)\,p_x\Bigr).$$

The symmetry group of point transformations can be computed using GEM package as well. The infinitesimal generators are given below:

$$\begin{aligned}
\xi_1 &= \mathscr{D}_t, \\
\xi_2 &= p\,\mathscr{D}_p, \\
\xi_3 &= -\frac{\mathrm{e}^{-\gamma t}}{\gamma}\,\mathscr{D}_t + \mathrm{e}^{-\gamma t}\,x\,\mathscr{D}_x - \mathrm{e}^{-\gamma t}\,p\,\mathscr{D}_p, \\
\xi_4 &= \frac{\mathrm{e}^{\gamma t}}{\gamma}\,\mathscr{D}_t + \mathrm{e}^{\gamma t}\,x\,\mathscr{D}_x - \frac{\gamma\,\mathrm{e}^{\gamma t}}{\eta}\,x\,p\,\mathscr{D}_p, \\
\xi_5 &= \mathrm{e}^{(\gamma + c)t}\,\mathcal{M}\Bigl(1 + \frac{c}{\gamma}, 1, \frac{\gamma x}{\eta}\Bigr)\,\mathrm{e}^{-\frac{\gamma x}{\eta}}\,\mathscr{D}_p, \\
\xi_6 &= \mathrm{e}^{(\gamma + c)t}\,\mathcal{U}\Bigl(1 + \frac{c}{\gamma}, 1, \frac{\gamma x}{\eta}\Bigr)\,\mathrm{e}^{-\frac{\gamma x}{\eta}}\,\mathscr{D}_p,
\end{aligned}$$

where $c \in \mathbb{R}$, $\mathcal{M}(a, b, z)$ and $\mathcal{U}(a, b, z)$ are KUMMER special functions [20,21] (see also Appendix C). The corresponding point transformations, which map solutions of (1) into other solutions, can be readily obtained by integrating several ODE systems (we do not provide integration details here):

$$\begin{gathered}
(t, x, p) \mapsto (t + \varepsilon_1, x, p), \\
(t, x, p) \mapsto (t, x, \mathrm{e}^{\varepsilon_2}\,p), \\
(t, x, p) \mapsto \Bigl(\frac{1}{\gamma}\ln(\varepsilon_3\gamma + \mathrm{e}^{\gamma t}), \frac{\mathrm{e}^{\gamma t}}{\varepsilon_3\gamma + \mathrm{e}^{\gamma t}}\,x, (1 + \varepsilon_3\gamma\,\mathrm{e}^{-\gamma t})\,p\Bigr), \\
(t, x, p) \mapsto \Bigl(t - \frac{1}{\gamma}\ln(1 - \varepsilon_4\gamma\mathrm{e}^{\gamma t}), \frac{x}{1 - \varepsilon_4\gamma\mathrm{e}^{\gamma t}}, \mathrm{e}^{-\frac{\varepsilon_4\gamma^2 x\,\mathrm{e}^{\gamma t}}{\eta(1 - \varepsilon_4\gamma\mathrm{e}^{\gamma t})}}\cdot p\Bigr), \\
(t, x, p) \mapsto \Bigl(t, x, p + \varepsilon_5\,\mathcal{M}\bigl(1 + \frac{c}{\gamma}, 1, \frac{\gamma x}{\eta}\bigr)\,\mathrm{e}^{-\frac{\gamma x}{\eta} + (\gamma + c)t}\Bigr), \\
(t, x, p) \mapsto \Bigl(t, x, p + \varepsilon_6\,\mathcal{U}\bigl(1 + \frac{c}{\gamma}, 1, \frac{\gamma x}{\eta}\bigr)\,\mathrm{e}^{-\frac{\gamma x}{\eta} + (\gamma + c)t}\Bigr).
\end{gathered}$$

The first symmetry is the time translation. The second one is the scaling of the dependent variable (the governing equation is linear). Symmetries 3 and 4 are exponential scalings. Two last symmetries express the fact that we can always add to the solution a particular solution to the homogeneous equation to obtain another solution. For instance, the solutions invariant under time translations ($\xi_1$) are steady states and their general form is the following:

$$p(x) = \mathrm{e}^{-\frac{\gamma x}{\eta}}\Bigl(\mathscr{C}_1\,\mathrm{E}_1\bigl(-\frac{\gamma x}{\eta}\bigr) + \mathscr{C}_2\Bigr), \tag{2}$$

where $\mathscr{C}_{1,2}$ are 'arbitrary' constants, which have to be determined from imposed conditions. Of course, they should be chosen so that the resulting steady solution is a valid probability distribution. It is not difficult to check that the imposed flux $\mathfrak{F}$ on the steady state solution is equal to $\mathscr{C}_1\,\eta$. Some properties of the exponential integral function are reminded in Appendix D.

We provide here also the general solutions invariant under the symmetry ($\xi_3$):

$$p(x, t) = \Bigl(\mathscr{C}_2 - \mathscr{C}_1\,t + \frac{\mathscr{C}_1}{\gamma}\ln x\Bigr)\,\mathrm{e}^{-\frac{\gamma x}{\eta}}$$

and under symmetry $(\xi_4)$:

$$p(x,\,t) \;=\; \Big(\mathscr{C}_1 + \mathscr{C}_2\, t + \frac{\mathscr{C}_2}{\gamma}\,\ln x\Big)\,\mathrm{e}^{\gamma t}.$$

These solutions might be used, for example, to validate numerical codes.

**Remark 1.** *As a byproduct of this analysis, we obtain two new exact solutions to the* FELLER *Equation* (1):

$$p(x,\,t) \;=\; \mathcal{M}\,\big(1 + \frac{c}{\gamma},\, 1,\, \frac{\gamma x}{\eta}\big)\,\mathrm{e}^{-\frac{\gamma x}{\eta} + (\gamma + c)\,t},$$
$$p(x,\,t) \;=\; \mathcal{U}\,\big(1 + \frac{c}{\gamma},\, 1,\, \frac{\gamma x}{\eta}\big)\,\mathrm{e}^{-\frac{\gamma x}{\eta} + (\gamma + c)\,t},$$

*for some constant* $c \in \mathbb{R}$.

## 3. Reformulation

By following the lines of ([17], Chapter 7), we are going to rewrite FELLER's Equation (1) with the so-called LAGRANGIAN or *material* variables. The main advantage of this formulation is due to the fact that we can handle infinite domains *without* any truncations, transformations, etc. It becomes possible to carry computations in infinite domains. Our domain is semi-infinite ($x \in \mathbb{R}^+$), with the left boundary $x = 0$ being a reflection point.

As the first step, we introduce the distribution function associated to the probability density $p(x,\,t)$:

$$\mathscr{P}(x,\,t) \;\stackrel{\mathrm{def}}{:=}\; \int_0^x p(\xi,\,t)\,\mathrm{d}\xi. \tag{3}$$

The same can be done for the initial condition as well:

$$\mathscr{P}_0(x) \;\stackrel{\mathrm{def}}{:=}\; \int_0^x p_0(\xi)\,\mathrm{d}\xi, \qquad p_0 \in W^{1,1}_{loc}(\mathbb{R}^+).$$

We notice also two obvious properties of the function $\mathscr{P}(x,\,t)$:

$$\partial_x \mathscr{P}(x,\,t) \equiv p(x,\,t), \qquad \lim_{x \to 0} \mathscr{P}(x,\,t) = 0, \qquad \lim_{x \to +\infty} \mathscr{P}(x,\,t) = 1.$$

Due to the positivity preservation property (see Appendix A), the function $\mathscr{P}(x,\,t)$ is nondecreasing in variable $x$. Thus, we can define its pseudo-inverse (This mapping is sometimes called in the literature as the *reciprocal mapping* [17] or an *order preserving string* [22].):

$$\mathscr{X}\,:\,[0,1]\times\mathbb{R}^+ \mapsto \mathbb{R}^+,$$

which can be computed as

$$\mathscr{X}(\bar{\mathscr{P}},\,t) \;\stackrel{\mathrm{def}}{:=}\; \inf\{\,\xi \in \mathbb{R}^+ \mid \mathscr{P}(\xi,\,t) = \bar{\mathscr{P}}\,\}.$$

The operation of taking the pseudo-inverse can be also seen as a generalized *hodograph transformation* $(x,\,t;\,\mathscr{P}) \mapsto (\mathscr{P},\,t;\,x)$ proposed presumably for the first time by Sir W.R. HAMILTON [23]. Similarly, the initial condition does possess a pseudo-inverse as well:

$$\mathscr{X}_0(\bar{\mathscr{P}}) \;\stackrel{\mathrm{def}}{:=}\; \inf\{\,\xi \in \mathbb{R}^+ \mid \mathscr{P}_0(\xi) = \bar{\mathscr{P}}\,\}, \tag{4}$$

such that $\mathscr{X}(\mathscr{P},\,0) \equiv \mathscr{X}_0(\mathscr{P})$.

If FELLER Equation (1) holds in the sense of distributions, then the following equation holds as well:

$$\mathscr{P}_t - x\left[\gamma\mathscr{P} + \eta\mathscr{P}_x\right]_x = 0. \tag{5}$$

along with the initial condition

$$\mathscr{P}(x, 0) = \mathscr{P}_0(x).$$

Equation (5) can be readily obtained by exploiting the obvious property $p(x, t) = \partial_x\mathscr{P}(x, t)$. In Appendix A, we show that zero value of the solution $p(x, t)$ is repulsive. Thus, $\partial_x\mathscr{P}(x, t) \equiv p(x, t) > 0,\ \forall(x, t) \in (\mathbb{R}^+)^2$. Thus, the implicit function theorem [24,25] guarantees the existence of derivatives of the inverse mapping $\mathscr{X}(\bar{\mathscr{P}}, t)$. Let us compute them by differentiating, with respect to $\bar{\mathscr{P}}$ and $t$, the following obvious identity:

$$\mathscr{P}(\mathscr{X}(\bar{\mathscr{P}}, t), t) \equiv \bar{\mathscr{P}}.$$

Thus, one can easily show that

$$\frac{\partial\mathscr{X}}{\partial\mathscr{P}} = \frac{1}{\partial_x\mathscr{P}}, \qquad \frac{\partial\mathscr{X}}{\partial t} = -\frac{\partial_t\mathscr{P}}{\partial_x\mathscr{P}}.$$

Using these expressions of partial derivatives, we derive the following evolution equation for the inverse mapping $\mathscr{X}(\mathscr{P}, \cdot)$:

$$\left(\mathrm{e}^{\gamma t}\mathscr{X}\right)_t + \mathscr{X}\cdot\left[\eta\,\mathrm{e}^{\gamma t}\left(\frac{\partial\mathscr{X}}{\partial\mathscr{P}}\right)^{-1}\right]_{\mathscr{P}} = 0. \tag{6}$$

The last equation can be rewritten also by introducing a new dynamic variable:

$$\mathscr{Y}(\mathscr{P}, t) \stackrel{\text{def}}{:=} \mathrm{e}^{\gamma t}\mathscr{X}(\mathscr{P}, t), \qquad \mathscr{Y}(\mathscr{P}, 0) \equiv \mathscr{X}(\mathscr{P}, 0). \tag{7}$$

It is not difficult to see that Equation (6) becomes

$$\mathscr{Y}_t + \mathscr{Y}\cdot\left[\eta\,\mathrm{e}^{\gamma t}\left(\frac{\partial\mathscr{Y}}{\partial\mathscr{P}}\right)^{-1}\right]_{\mathscr{P}} = 0. \tag{8}$$

The last equation will be solved numerically in the following Section.

**Remark 2.** *We would like to underline the fact that, by our assumptions,* $\frac{\partial\mathscr{Y}}{\partial\mathscr{P}}$ *as well as* $\frac{\partial\mathscr{X}}{\partial\mathscr{P}}$ *cannot vanish. Thus, there is no problem in dividing by* $\frac{\partial\mathscr{Y}}{\partial\mathscr{P}}$ *in Equation* (8).

### 3.1. Numerical Discretization

Earlier, we derived Equation (6), which governs the dynamics of the pseudo-inverse mapping $\mathscr{X}(\mathscr{P}, \cdot)$. The initial condition for Equation (6) is given by the pseudo-inverse (4) of the initial condition $\mathscr{P}_0(x)$. We discretize Equation (6) with an explicit discretization in time since it yields the most straightforward implementation.

The first step in our algorithm consists of choosing the initial sampling interval. We make this choice depending on the provided initial condition. Typically, we want to sample only where it is needed. Thus, it seems reasonable to choose the *initial* segment $[0, \ell_0]$, with $\ell_0$ being the leftmost location such that

$$1 - \mathscr{P}_0(\ell_0) < \mathsf{tol}.$$

In simulations presented below, we chose $\mathsf{tol} \sim \mathcal{O}(10^{-5})$. Then, we chose the initial sampling $\{\mathscr{X}_k^0\}_{k=0}^N \in [0, \ell_0] \subseteq \mathbb{R}^+$, with $\mathscr{X}_0^0 = 0$ and $\mathscr{X}_N^0 = \ell_0$. It is desirable that the initial sampling be adapted to the initial condition, since errors made initially cannot be corrected later. One of the possible strategies for the initial grid generation can be found in ([26], Section 2.3.1). We define also $\mathscr{P}_k = \mathscr{P}_0(\mathscr{X}_k^0)$. We stress that $\{\mathscr{P}_k\}_{k=0}^N$ stand for a discrete cumulative mass variable and, thus, they are time independent.

More generally, we introduce the following notation:

$$\mathscr{X}_k^n \stackrel{\text{def}}{:=} \mathscr{X}(\mathscr{P}_k, t^n), \qquad k = 0, 1, \ldots, N,$$

with $t^n \stackrel{\text{def}}{:=} n\,\Delta t$, $n \in \mathbb{N}$ and $\Delta t > 0$ being a chosen time step. (We present our algorithm with a constant time step for the sake of simplicity. However, in realistic simulations presented in Section 4, the time step will be chosen adaptively and automatically to meet the stability and accuracy requirements prescribed by the user.) We introduce also similar notation for the dynamic variable:

$$\mathscr{Y}_k^n \stackrel{\text{def}}{:=} \mathscr{Y}(\mathscr{P}_k, t^n), \qquad \mathscr{Y}_k^0 \equiv \mathscr{X}_k^0, \qquad k = 0, 1, \ldots, N.$$

Now, we can state the fully discrete scheme for Equation (8):

$$\frac{\mathscr{Y}_k^{n+1} - \mathscr{Y}_k^n}{\Delta t} + \eta\, \mathrm{e}^{\gamma t^n} \frac{\mathscr{Y}_k^n}{\Delta \mathscr{P}_k} \left\{ \frac{\Delta \mathscr{P}_{k+\frac{1}{2}}}{\mathscr{Y}_{k+1}^n - \mathscr{Y}_k^n} - \frac{\Delta \mathscr{P}_{k-\frac{1}{2}}}{\mathscr{Y}_k^n - \mathscr{Y}_{k-1}^n} \right\}, \tag{9}$$

with $n \geqslant 0$, $k = 0, 1, \ldots, N-1$ and

$$\Delta \mathscr{P}_{k+\frac{1}{2}} \stackrel{\text{def}}{:=} \mathscr{P}_{k+1} - \mathscr{P}_k, \qquad \Delta \mathscr{P}_{k-\frac{1}{2}} \stackrel{\text{def}}{:=} \mathscr{P}_k - \mathscr{P}_{k-1}.$$

The quantity $\Delta \mathscr{P}_k$ can be defined as the arithmetic or geometric mean of two neighbouring discretization steps $\Delta \mathscr{P}_{k \pm \frac{1}{2}}$:

$$\Delta \mathscr{P}_k \stackrel{\text{def}}{:=} \frac{\Delta \mathscr{P}_{k+\frac{1}{2}} + \Delta \mathscr{P}_{k-\frac{1}{2}}}{2}, \qquad \Delta \mathscr{P}_k \stackrel{\text{def}}{:=} \sqrt{\Delta \mathscr{P}_{k+\frac{1}{2}} \cdot \Delta \mathscr{P}_{k-\frac{1}{2}}}.$$

To be specific, in our code, we implemented the arithmetic mean. The fully discrete scheme can be easily rewritten under the form of a discrete dynamical system:

$$\mathscr{Y}_k^{n+1} = \mathscr{Y}_k^n - \eta\, \Delta t\, \mathrm{e}^{\gamma t^n} \frac{\mathscr{Y}_k^n}{\Delta \mathscr{P}_k} \left\{ \frac{\Delta \mathscr{P}_{k+\frac{1}{2}}}{\mathscr{Y}_{k+1}^n - \mathscr{Y}_k^n} - \frac{\Delta \mathscr{P}_{k-\frac{1}{2}}}{\mathscr{Y}_k^n - \mathscr{Y}_{k-1}^n} \right\}, \quad n \geqslant 0.$$

**Remark 3.** *We would like to say a few words about the implementation of boundary conditions. First of all, no boundary condition is required on the left side, where $\mathscr{X}_0^n = \mathscr{Y}_0^n \equiv 0$. On the right boundary, we prefer to impose the infinite* NEUMANN*-type boundary condition (i.e., $\frac{\partial \mathscr{Y}}{\partial \mathscr{P}} \to \infty$), which yields the exact 'mass' conservation at the discrete level as well. Namely, at the rightmost cell, we have the following fully discrete scheme:*

$$\mathscr{Y}_N^{n+1} = \mathscr{Y}_N^n + \eta\, \Delta t\, \mathrm{e}^{\gamma t^n} \frac{\mathscr{Y}_N^n}{\Delta \mathscr{P}_N} \cdot \frac{\Delta \mathscr{P}_{N-\frac{1}{2}}}{\mathscr{Y}_N^n - \mathscr{Y}_{N-1}^n}, \qquad n \geqslant 0,$$

*with $\Delta \mathscr{P}_N \stackrel{\text{def}}{:=} \dfrac{\mathscr{P}_N - \mathscr{P}_{N-2}}{2}$. As a result, we obtain the exact conservation of 'mass' at the discrete level:*

$$\sum_{k=0}^{N} \Delta \mathscr{P}_k\, \mathscr{X}_k^n \equiv \sum_{k=0}^{N} \Delta \mathscr{P}_k\, \mathscr{X}_k^0, \qquad \forall n \in \mathbb{N}.$$

To summarize, our numerical strategy consists of the following steps:

1. We compute the pseudo-inverse of the initial data $p_0(x)$ to obtain $\mathscr{X}(\mathscr{P}, 0) \equiv \mathscr{Y}(\mathscr{P}, 0)$.
2. This initial condition $\mathscr{Y}(\mathscr{P}, 0)$ is evolved in (discrete) time using an *explicit* marching scheme in order to obtain numerical approximation to $\mathscr{Y}(\mathscr{P}, t)$, $t > 0$.
3. The variable $\mathscr{X}(\mathscr{P}, t)$ is recovered by inverting (7), i.e.,

$$\mathscr{X}(\mathscr{P}, t) = \mathrm{e}^{-\gamma t}\,\mathscr{Y}(\mathscr{P}, t).$$

4. Thanks to (3), we can deduce the values of $p\left(\mathscr{X}(\mathscr{P}, t), t\right) \in [0, 1]$ by applying a favorite finite difference formula (In our code, we employed the simplest forward finite differences, and it led satisfactory results. This point can be easily improved when necessary.).

Working with the pseudo-inverse allows us to overcome the issue of the retention phenomenon, which manifests as the expanding support of $p(x, t)$ for positive (and possibly large) times $t > 0$, $t \gg 1$, since the computational domain was transformed to $[0, 1]$. This method is the LAGRANGIAN counterpart of the moving mesh technique in the EULERIAN setting [26,27].

A simple MATLAB code, which implements the scheme we described above, is freely available for reader's convenience at [28].

## 4. Numerical Results

In this Section, we validate and illustrate the application of the proposed algorithm on several examples. However, first, we begin with a straightforward validation test. The only difference with the proposed algorithm above is that we are using a higher-order adaptive time stepping for our practical simulations. The explicit first-order scheme was used to simplify the presentation. In practice, much more sophisticated time steppers can be used. For instance, we shall employ the explicit embedded DORMAND–PRINCE RUNGE–KUTTA pair (4, 5) [29] implemented in MATLAB under the `ode45` routine [30]. Conceptually, this method is similar to the explicit EULER scheme presented above. It conserved all good properties we showed, but it provides additionally the higher accuracy order and totally automatic adaptivity of the time step, which matches very well with adaptive features of the LAGRANGIAN scheme in space. The values of absolute and relative tolerances used in the time step choice are systematically reported below.

### 4.1. Steady State Preservation

In order to validate the numerical algorithm, we have at our disposal a family of steady state solutions (2). Hence, if we take such a solution as an initial condition, normally the algorithm has to keep it intact under the discretized dynamics. The parameters $\eta$, $\gamma$ of the equation, those of the steady solution, and numerical parameters used in our computation are reported in Table 1. The initial condition at $t = 0$ along with the final state at $t = T$ are shown in Figure 1. Up to graphical resolution they coincide completely. We can easily check that during the whole simulation the points barely moved, i.e.,

$$\|\, \mathscr{X}(\cdot, T) - \mathscr{X}(\cdot, 0) \,\|_\infty \approx 0.008577\ldots$$

We can check also other quantities. For instance, $\mathscr{P}(\xi, t)$ is preserved up to the machine precision. If we reconstruct the probability distribution $p(x, t)$, we obtain

$$\|\, p(\cdot, T) - p(\cdot, 0) \,\|_\infty \approx 0.003051\ldots$$

The last error comes essentially from the fact that we apply simple first-order finite difference to reconstruct the variable $p(x, t)$ from its primitive $\mathscr{P}(x, t)$. We can improve this point, but even this simple reconstruction seems to be consistent with the overall scheme accuracy. Thus, this example shows that our implementation of the proposed algorithm is also practically well-balanced [17], since steady state solutions are well preserved.

**Table 1.** Numerical parameters used in the steady state computations.

| Parameter | Value |
|---|---|
| Drift coefficient, $\gamma$ | 1.0 |
| Diffusion coefficient, $\eta$ | 1.0 |
| Integration constant, $\mathscr{C}_1$ | 0.0 |
| Integration constant, $\mathscr{C}_2$ | $\frac{\eta}{\gamma} \equiv 1.0$ |
| Final simulation time, $T$ | 10.0 |
| Number of discretization points, $N$ | 500 |
| Absolute tolerance, $\mathsf{tol}_{\,\mathrm{a}}$ | $10^{-5}$ |
| Relative tolerance, $\mathsf{tol}_{\,\mathrm{r}}$ | $10^{-5}$ |

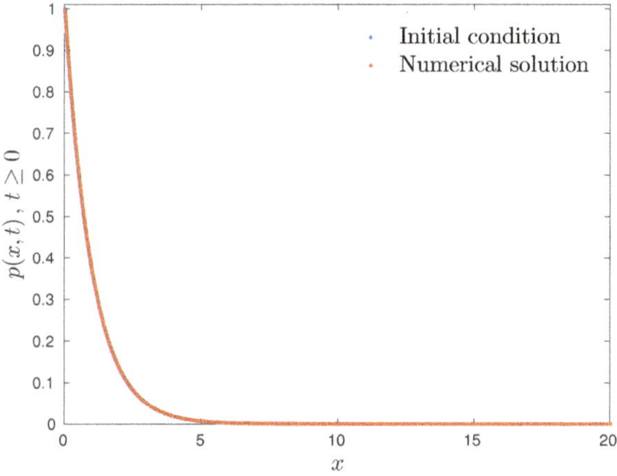


**Figure 1.** Comparison of a steady state solution of class (2) at $t = 0$ and at $t = T = 10$. They are indistinguishable up to the graphical resolution, which validates the solver.

## 4.2. Transient Computations

In this Section, we present a couple of extra truly unsteady computations in order to illustrate the capabilities of our method. Namely, we shall simulate the probability distributions emerging from a family of initial conditions (normalized to have the following probability distribution):

$$p_0(x) = \frac{e^{-\frac{x}{\sigma_1}} + e^{-\frac{x - x_0}{\sigma_2}}}{\sigma_1 + \sigma_2 e^{\frac{x_0}{\sigma_2}}}.$$

The primitive of the last distribution can be easily computed as well:

$$\mathscr{P}(x) = 1 - \frac{\sigma_1 e^{-\frac{x}{\sigma_1}} + \sigma_2 e^{-\frac{x - x_0}{\sigma_2}}}{\sigma_1 + \sigma_2 e^{\frac{x_0}{\sigma_2}}}.$$

We design two different experiments in silico to show two completely different behaviour of solutions to FELLER Equation (1) depending on the sign of the drift coefficient $\gamma$. These will constitute additional tests for the proposed numerical method. In both cases, the initial positions of particles are chosen according to the logarithmic distribution (`logspace` function in MATLAB) on the segment $[0, 20]$. This choice is made to represent more accurately the exponentially decaying initial condition since the errors made in the initial condition cannot be corrected in the dynamics.

4.2.1. Expanding Drift

If the drift coefficient $\gamma < 0$, this term will cause FELLER's Equation (1) to transport information in the rightward direction. This situation is quite uncomfortable from the numerical point of view since the initial condition expands quickly towards the (positive) infinity. We submitted our method to this case. All numerical and physical parameters are provided in Table 2 (the middle column). The initial condition along with the probability distribution at the end of our simulation are shown simultaneously in Figure 2a,b in CARTESIAN and semilogarithmic coordinates correspondingly. As expected, the support of the initial condition more than triples from $t = 0$ to $t = T = 3.0$. We remind that the diffusion and drift coefficients are proportional to $x$, and the scheme handles the growth of these terms automatically. The smooth decay of the solution on the semilogarithmic plot (see Figure 2b) shows the absence of any numerical instabilities. The trajectory of grid nodes is shown in Figure 2c. One can see that points follow the expansion of the solution. Nevertheless, they concentrate in the areas where the probability takes significant values.

**Table 2.** Numerical parameters used in unsteady computations.

| Parameter | Expanding Experiment | Confining Experiment |
|---|---|---|
| Drift coefficient, $\gamma$ | $-0.1$ | 0.5 |
| Diffusion coefficient, $\eta$ | 1.0 | 1.0 |
| Final simulation time, $T$ | 3.0 | 12.0 |
| Number of discretization points, $N$ | 100 | 100 |
| Initial condition parameter, $\sigma_1$ | 2.0 | 2.0 |
| Initial condition parameter, $\sigma_2$ | 1.0 | 1.0 |
| Initial condition parameter, $x_0$ | 3.0 | 3.0 |
| Absolute tolerance, $\mathrm{tol}_a$ | $10^{-5}$ | $10^{-5}$ |
| Relative tolerance, $\mathrm{tol}_r$ | $10^{-5}$ | $10^{-5}$ |

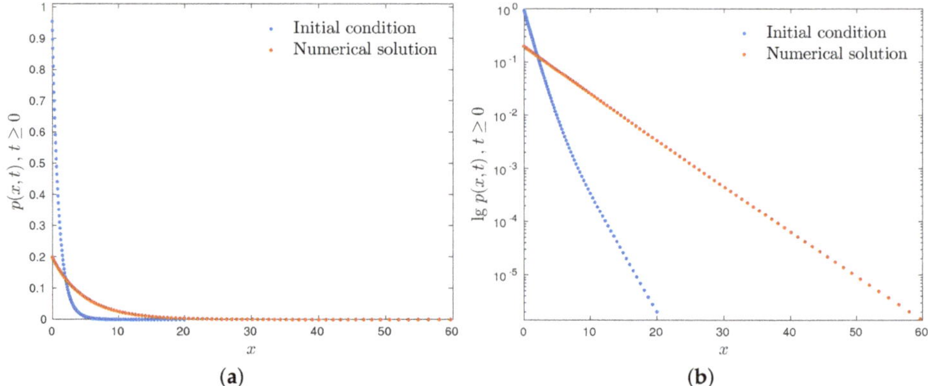


(a) (b)

**Figure 2.** *Cont.*

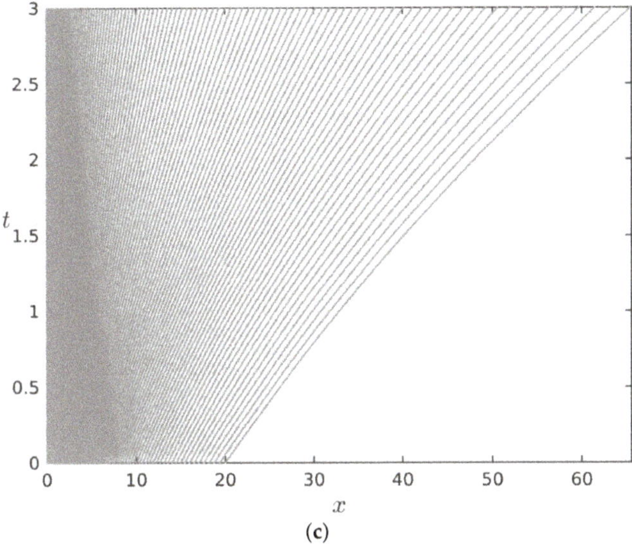


(c)

**Figure 2.** Numerical result of the expanding experiment with negative drift $\gamma = -0.1 < 0$: (**a**) initial and terminal states of the numerical discretized solution; (**b**) initial and terminal states of the numerical discretized solution in semilogarithmic coordinates; and (**c**) trajectories of grid nodes. All numerical parameters for this computation are reported in Table 2 (middle column).

#### 4.2.2. Confining Drift

In the case of the positive drift coefficient $\gamma > 0$, the FELLER dynamics get even more interesting since we have two competing effects:

1. Positive drift moving information towards the origin $x = 0$;
2. Diffusion spreading solution towards the positive infinity.

It might happen that both effects balance each other and result in a steady solution. Such a simulation is presented in this Section. The numerical and physical parameters are given in Table 2 (the right column). The initial condition along with the probability distribution at the end of our simulations are shown simultaneously in Figure 3a,b in CARTESIAN and semilogarithmic coordinates, respectively. The trajectories of grid nodes are shown in Figure 2c. One can see the rapid initial expansion stage, which is followed by a stabilization process, when we almost achieved the expected steady state. Again, the grid nodes trajectories show excellent adaptive features of the numerical scheme—at the end of the simulation, the relative density of nodes turns out to be preserved. The semilogarithmic plot shown in Figure 3b shows that the numerical solution is free of numerical instabilities. The obtained steady solution will be preserved by discrete dynamics thanks to the well-balanced property demonstrated in Section 4.1.

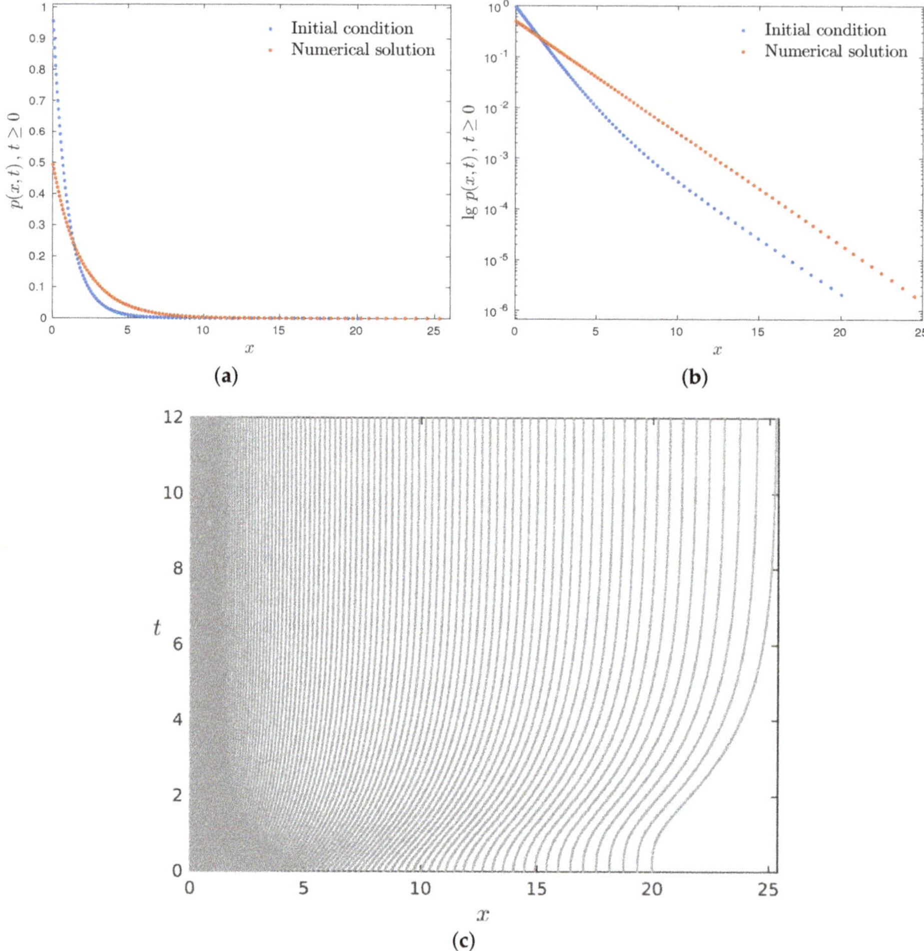


**Figure 3.** Numerical result of the confining experiment with positive drift $\gamma = 0.5 > 0$: (**a**) initial and terminal states of the numerical discretized solution; (**b**) initial and terminal states of the numerical discretized solution in semilogarithmic coordinates; and (**c**) trajectories of grid nodes. All numerical parameters for this computation are reported in Table 2 (right column).

## 5. Discussion

Above, we presented some rationale about the discretization, existence, and uniqueness theory for FELLER's equation. The main conclusions and perspectives are outlined below.

### 5.1. Conclusions

The celebrated FELLER equation was studied mathematically in two seminal papers published by William FELLER (1951/1952) in Annals of Mathematics [1,2]. In particular, the uniqueness was shown in [2] without any boundary condition required at $x = 0$. This result is notable and we use it in our study. The main goal of our work was to present an efficient numerical scheme, which is able to handle the unbounded diffusion inherent to Equation (1). Moreover, the retention phenomenon causes the solution support to expand all the time. Thus, if we wait a sufficiently long time, it will reach the computational domain boundaries (since $\mathbb{R}^+ \ni x$ was truncated to a finite segment to make the in silico simulation possible). To overcome this difficulty, we proposed a simple and explicit LAGRANGIAN scheme using the notion of the pseudo-inverse. In this way, we obtained a stable numerical scheme (under not so stringent stability conditions), which turns out to be naturally *adaptive* as well, since particles move together with the growing support (the rightmost particle position defines the support) and tend to concentrate where it is really needed. The performance of our scheme was illustrated in several examples. We share also the MATLAB code so that anybody can check the claims we made above and use it to solve numerically the FELLER equation in their applications: https://github.com/dutykh/Feller/.

### 5.2. Perspectives

All the numerical schemes and results presented in this paper were in one 'spatial' dimension. The FELLER equation considered here is 1D as well. However, it seems interesting (regardless of the physical applications) to consider generalized FELLER equations in higher space dimensions and to extend the proposed numerical strategy to the multidimensional case as well. Another possible extension direction consists of proposing higher-order schemes to capture numerical solutions with better accuracy with the same number of degrees of freedom. On the more theoretical side, we would like to obtain an alternative well-posedness theory for FELLER equation by taking a continuous limit in our numerical scheme.

**Funding:** This research was not supported by any external funding.

**Acknowledgments:** The author would like to thank Prof. Laurent GOSSE (CNR–IAC Rome, Italy) for his invaluable help in understanding his first book. I thank also Prof. Francesco FEDELE (Georgia Tech, Atlanta, USA) for pointing out to me this interesting problem.

**Conflicts of Interest:** There is no conflict of interests.

## Abbreviations

The following abbreviations are used in this manuscript:

| | |
|---|---|
| CFL | Courant Friedrichs Lew |
| GeM | A Maple module for the computation of symmetries and conservation laws of differential equations |
| ODE | Ordinary Differential Equation |
| PDF | Probability Density Function |
| PDE | Partial Differential Equation |
| WT | Wave Turbulence |

## Appendix A. Positivity Preservation

In this Appendix, we show the positivity preservation property of the solution $p(x, t)$ to Equation (1). We suppose that initially the solution is non-negative, i.e.,

$$p(x, 0) = p_0(x) \geqslant 0, \qquad \forall x \in \mathbb{R}^+.$$

Let us assume that at some positive time $t = t^* > 0$ and in some point $x = x^* \in \mathbb{R}^+$ the solution attains zero value, i.e.,

$$p(x^*, t^*) = 0.$$

This situation is schematically depicted in Figure A1.

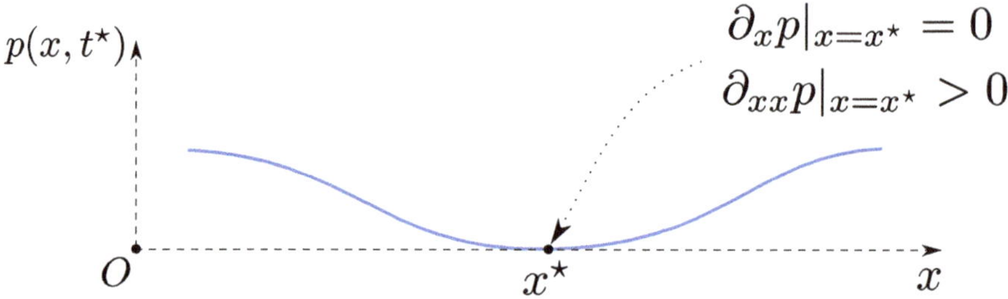


**Figure A1.** A schematic representation of the situation where the solution $p(x, t)$ attains a zero value at some point $x^\star > 0$.

Equation (1) can be rewritten in the nonconservative form:

$$p_t = x\left[\gamma p_x + \eta p_{xx}\right] + \gamma p + \eta p_x .$$

Taking into account the fact that the point $x^\star$ is a local minimum ($p_x|_{x^\star} = 0$), where the solution takes zero value ($p|_{x^\star} = 0$), the last equation greatly simplifies at this point:

$$p_t|_{x^\star} = x^\star \eta p_{xx}|_{x^\star} > 0,$$

since in the minimum $p_{xx}|_{x^\star} > 0$. Thus, zero values of the solution are *repulsive* and for (at least small) times $t > t^\star$, the function $t \mapsto p(x^\star, t)$ will be increasing.

## Appendix B. 'Mass' conservation

It is straightforward to show that the $L_1$ norm of the solution to Equations (1) is preserved. Indeed, taking into account that the solution $p(x, t) > 0$ is positive for all times $t > 0$ provided that the initial condition $p_0(x) \geqslant 0, \forall x \in \mathbb{R}^+$, we have $|p(x, t)| \equiv p(x, t)$. By integrating Equation (1), we have

$$\partial_t \int_{\mathbb{R}^+} p(x, t)\,\mathrm{d}x + \int_{\mathbb{R}^+} \mathfrak{F}_x\,\mathrm{d}x = 0 .$$

Taking into account that $\mathfrak{F}|_{x=0} \equiv 0$ and the solution $p(x, t)$ is decaying sufficiently fast, as $x \to +\infty$ together with its derivative, we obtain

$$\partial_t \int_{\mathbb{R}^+} p(x, t)\,\mathrm{d}x \equiv 0 .$$

In other words,

$$\|p(\cdot, t)\|_{L_1} = \text{const} .$$

The last constant can be in general taken equal to one after the appropriate rescaling (provided that the initial condition is nontrivial). It is this scaling which is assumed throughout the whole text above. This renormalization is consistent, with the 'physical sense' of the variable $p$ being the density of a probability distribution.

## Appendix C. Kummer's Functions

The KUMMER functions $\mathcal{M}(a, b, z)$ and $\mathcal{U}(a, b, z)$ are two linearly independent solutions of the following ordinary differential equation:

$$z\frac{\mathrm{d}y^2}{\mathrm{d}z^2} + (b - z)\frac{\mathrm{d}y}{\mathrm{d}z} - a y = 0 .$$

This equation admits two singular points: $z = 0$ (regular) and $z = +\infty$ (irregular). There exists a connection between KUMMER and hypergeometric functions [20].

## Appendix D. Exponential Integral

Notice also an important property of the exponential integral function $\mathrm{E}_1(x)$, which is useful in manipulating its values for negative arguments:

$$\lim_{\delta \to 0} \mathrm{E}_1(x \pm \mathrm{i}\delta) = \mathrm{E}_1(x) \mp \mathrm{i}\pi \equiv -\mathrm{Ei}(-x) \mp \mathrm{i}\pi, \qquad x < 0,$$

where $\mathrm{Ei}(z)$ is the following exponential integral:

$$\mathrm{Ei}(z) \stackrel{\mathrm{def}}{:=} -\int_{-\infty}^{z} \frac{\mathrm{e}^t}{t}\,\mathrm{d}t.$$

Thus, $\mathrm{Ei}(x) \equiv -\mathrm{E}_1(-x)$ for $x < 0$. The last relation can be also extended to the entire complex plain as follows:

$$\mathrm{Ei}(z) \equiv -\mathrm{E}_1(-z) + \frac{1}{2}\ln z - \frac{1}{2}\ln\left(\frac{1}{z}\right) - \ln(-z), \qquad z \in \mathbb{C}.$$

More properties of the exponential integral can be found in the specialized literature (see e.g., ([20], Chapter 5)).

## References

1. Feller, W. Two Singular Diffusion Problems. *Ann. Math.* **1951**, *54*, 173. [CrossRef]
2. Feller, W. The Parabolic Differential Equations and the Associated Semi-Groups of Transformations. *Ann. Math.* **1952**, *55*, 468. [CrossRef]
3. Masoliver, J. Nonstationary Feller process with time-varying coefficients. *Phys. Rev. E* **2016**, *93*, 012122. [CrossRef] [PubMed]
4. Risken, H.; Till, F. *The Fokker-Planck Equation: Methods of Solution and Applications*, 2nd ed.; Springer Series in Synergetics; Springer: Berlin/Heidelberg, Germany, 1996; Volume 18.
5. Gardiner, C. *Handbook of Stochastic Methods: for Physics, Chemistry and the Natural Sciences*, 3rd ed.; Springer: Berlin/Heidelberg, Germany, 2004.
6. Gan, X.; Waxman, D. Singular solution of the Feller diffusion equation via a spectral decomposition. *Phys. Rev. E* **2015**, *91*, 012123. [CrossRef] [PubMed]
7. Murray, J.D. *Mathematical Biology: I. An Introduction*; Springer: Berlin/Heidelberg, Germany, 2007.
8. Lehnikg, S.H. *The Generalized Feller Equation and Related Topics*; Longman Scientific & Technical: Harlow, Essex, UK, 1993.
9. McKinstrie, C.; Lakoba, T. Probability-density function for energy perturbations of isolated optical pulses. *Opt. Express* **2003**, *11*, 3628–3648. [CrossRef] [PubMed]
10. Choi, Y.; Lvov, Y.V.; Nazarenko, S.; Pokorni, B. Anomalous probability of large amplitudes in wave turbulence. *Phys. Lett. A* **2005**, *339*, 361–369. [CrossRef]
11. Zakharov, V.E.; Lvov, V.S.; Falkovich, G. *Kolmogorov Spectra of Turbulence I. Wave Turbulence*; Series in Nonlinear Dynamics; Springer: Berlin, Germany, 1992.
12. Nazarenko, S. *Wave Turbulence*; Springer Series Lecture Notes in Phyics; Springer: Berlin, Germany, 2011; Volume 825.
13. Choi, Y.; Jo, S.; Kwon, Y.-S.; Nazarenko, S. Nonstationary distributions of wave intensities in wave turbulence. *J. Phys. A Math. Gen.* **2017**, *50*, 355502. [CrossRef]
14. Dysthe, K.B.; Krogstad, H.E.; Muller, P. Oceanic rogue waves. *Ann. Rev. Fluid Mech.* **2008**, *40*, 287–310. [CrossRef]
15. Courant, R.; Friedrichs, K.; Lewy, H. Über die partiellen Differenzengleichungen der mathematischen Physik. *Math. Ann.* **1928**, *100*, 32–74. [CrossRef]

16. Carrillo, J.A.; Gualdani, M.P.; Toscani, G. Finite speed of propagation in porous media by mass transportation methods. *C. R. Mathématique* **2004**, *338*, 15–818. [CrossRef]
17. Gosse, L. *Computing Qualitatively Correct Approximations of Balance Laws: Exponential-Fit, Well-Balanced and Asymptotic-Preserving*; 1st ed.; Volume 2 of SIMAI Springer Series; Springer: Milan, Italy, 2013.
18. Bluman, G.W.; Cheviakov, A.F.; Anco, S.C. *Applications of Symmetry Methods to Partial Differential Equations*; Springer: New York, NY, USA, 2010.
19. Cheviakov, A.F. GeM software package for computation of symmetries and conservation laws of differential equations. *Comput. Phys. Comm.* **2007**, *176*, 48–61. [CrossRef]
20. Abramowitz, M.; Stegun, I.A. *Handbook of Mathematical Functions*; Dover Publications: Mineola, NY, USA, 1972.
21. Kummer, E.E. De integralibus quibusdam definitis et seriebus infinitis. *J. Für Die Reine Und Angew. Math.* **1837**, *17*, 228–242.
22. Brenier, Y. Order preserving vibrating strings and applications to electrodynamics and magnetohydrodynamics. *Methods Appl. Anal.* **2004**, *11*, 512–532. [CrossRef]
23. Hamilton, R.W. The Hodograph, or a New Method of Expressing in Symbolic Language the Newtonian Law of Attraction. *Proc. R. Irish Acad.* **1847**, *3*, 344–353.
24. Zorich, V.A. *Mathematical Analysis I*, 2nd ed.; Springer Verlag: Berlin/Heidelberg, Germany; New York, NY, USA, 2008.
25. Zorich, V.A. *Mathematical Analysis II*, 2nd ed.; Springer Verlag: Berlin/Heidelberg, Germany; New York, NY, USA, 2008.
26. Khakimzyanov, G.S.; Dutykh, D.; Mitsotakis, D.; Shokina, N.Y. Numerical simulation of conservation laws with moving grid nodes: Application to tsunami wave modelling. *Geosciences* **2019**, *9*, 197. [CrossRef]
27. Khakimzyanov, G.; Dutykh, D. On supraconvergence phenomenon for second order centered finite differences on non-uniform grids. *J. Comput. Appl. Math.* **2017**, *326*, 1–14. [CrossRef]
28. Dutykh, D. 2019. Available online: https://github.com/dutykh/Feller/ (accessed on 1 November 2019).
29. Dormand, J.R.; Prince, P.J. A family of embedded Runge-Kutta formulae. *J. Comput. Appl. Math.* **1980**, *6*, 19–26. [CrossRef]
30. Shampine, L.F.; Reichelt, M.W. The MATLAB ODE Suite. *SIAM J. Sci. Comput.* **1997**, *18*, 1–22. [CrossRef]

Finally, (9) was deduced in [34] in the context of plasma physic and that similar equations describe the dynamics of radiating gases [35,36], in [37–40] in the context of ultrafast pulse propagation in a mode-locked laser cavity in the few-femtosecond pulse regime and in [41] in the context of Maxwell equations.

The Cauchy problem for (9) was studied in [42–44] in the context of energy spaces, [4,5,45,46] in the context of entropy solutions. The homogeneous initial boundary value problem was studied in [47–50]. Nonlocal formulations of (9) were analyzed in [15,51] and the convergence of a finite difference scheme proved in [52].

Observe that, taking $\alpha = \beta = 0$, (1) reads

$$\begin{cases} \partial_t u + \left(g + \frac{q\kappa}{\gamma}\right)\partial_x u^3 - a\partial_x^3 u = bP, & t > 0,\ x \in \mathbb{R}, \\ \partial_x P = u, & t > 0,\ x \in \mathbb{R}. \end{cases} \tag{10}$$

It was derived by Costanzino, Manukian and Jones [53] in the context of the nonlinear Maxwell equations with high-frequency dispersion. Kozlov and Sazonov [12] show that (10) is an more general equation than (9) to describe the nonlinear propagation of optical pulses of a few oscillations duration in dielectric media.

Mathematical properties of (10) are studied in many different contexts, including the local and global well-posedness in energy spaces [43,53] and stability of solitary waves [53,54], while, in [6], the well-posedness of the classical solutions is proven.

Observe that letting $a \to 0$ in (10), we obtain (9). Hence, following [19,55,56], in [5,57], the convergence of the solution of (10) to the unique entropy solution of (9).

The main result of this paper is the following theorem.

**Theorem 1.** *Assume (2), (3), (4) and (5). Fix $T > 0$, there exists an unique solution $(u, v, P)$ of (1) such that*

$$\begin{aligned} &u \in L^\infty(0,T;H^2(\mathbb{R})), \\ &v \in L^\infty(0,T;H^4(\mathbb{R})), \\ &P \in L^\infty(0,T;H^3(\mathbb{R})). \end{aligned} \tag{11}$$

*In particular, we have that*

$$\int_{\mathbb{R}} u(t,x)dx = 0, \quad t \geq 0. \tag{12}$$

*Moreover, if $(u_1, v_1, P_1)$ and $(u_2, v_2, P_2)$ are two solutions of (1), we have that*

$$\begin{aligned} \|u_1(t,\cdot) - u_2(t,\cdot)\|_{L^2(\mathbb{R})} &\leq e^{C(T)t}\|u_{1,0} - u_{2,0}\|_{L^2(\mathbb{R})}, \\ \|v_1(t,\cdot) - v_2(t,\cdot)\|_{H^2(\mathbb{R})} &\leq e^{C(T)t}\|u_{1,0} - u_{2,0}\|_{L^2(\mathbb{R})}, \\ \|P_1(t,\cdot) - P_2(t,\cdot)\|_{H^1(\mathbb{R})} &\leq e^{C(T)t}\|P_{1,0} - P_{2,0}\|_{H^1(\mathbb{R})}, \end{aligned} \tag{13}$$

*where,*

$$P_{1,0}(x) = \int_{-\infty}^x u_{1,0}(y)dy, \quad P_{2,0}(x) = \int_{-\infty}^x u_{2,0}(y)dy, \tag{14}$$

*for some suitable $C(T) > 0$, and every $0 \leq t \leq T$.*

The proof of Theorem 1 is based on the Aubin–Lions Lemma (see [58–60]).

The paper is organized as follows. In Section 2, we prove several a priori estimates on a vanishing viscosity approximation of (1). Those play a key role in the proof of our main result, that is given in

Section 3. Appendix A is an appendix, where we prove the posedness of the classical solutions of (1), under the assumption

$$u_0 \in L^1(\mathbb{R}) \cap H^3(\mathbb{R}). \tag{15}$$

## 2. Vanishing Viscosity Approximation

Our existence argument is based on passing to the limit in a vanishing viscosity approximation of (1).

Fix a small number $0 < \varepsilon < 1$ and let $u_\varepsilon = u_\varepsilon(t, x)$ be the unique classical solution of the following mixed problem [19,61,62]:

$$\begin{cases} \partial_t u_\varepsilon + 3gu_\varepsilon^2 \partial_x u_\varepsilon - a\partial_x^3 u_\varepsilon + q\partial_x(v_\varepsilon u_\varepsilon) = bP_\varepsilon - \varepsilon\partial_x^4 u_\varepsilon, & t > 0,\ x \in \mathbb{R}, \\ \partial_x P_\varepsilon = u_\varepsilon, & t > 0,\ x \in \mathbb{R}, \\ \alpha\partial_x^2 v_\varepsilon + \beta\partial_x v_\varepsilon + \gamma v_\varepsilon = \kappa u_\varepsilon^2, & t > 0,\ x \in \mathbb{R}, \\ P_\varepsilon(t, -\infty) = 0, & t > 0, \\ u_\varepsilon(0, x) = u_{\varepsilon,0}(x), & x \in \mathbb{R}, \end{cases} \tag{16}$$

where $u_{\varepsilon,0}$ is a $C^\infty$ approximation of $u_0$ such that

$$\begin{aligned} &\|u_{\varepsilon,0}\|_{H^2(\mathbb{R})} \le \|u_0\|_{H^2(\mathbb{R})}, \quad \int_{\mathbb{R}} u_{\varepsilon,0} dx = 0, \\ &\|P_{\varepsilon,0}\|_{L^2(\mathbb{R})} \le \|P_0\|_{L^2(\mathbb{R})}, \quad \int_{\mathbb{R}} P_{\varepsilon,0} dx = 0. \end{aligned} \tag{17}$$

Let us prove some a priori estimates on $u_\varepsilon$, $P_\varepsilon$ and $v_\varepsilon$. We denote with $C_0$ the constants which depend only on the initial data, and with $C(T)$, the constants which depend also on $T$.

**Lemma 1.** *For each $t \ge 0$,*

$$P_\varepsilon(t, \infty) = 0. \tag{18}$$

*In particular, we have that*

$$\int_{\mathbb{R}} u_\varepsilon(t, x) dx = 0. \tag{19}$$

**Proof.** We begin by proving (18). Thanks to the regularity of $u_\varepsilon$ and the first equation of (16), we have that

$$\lim_{x \to \infty} \left( \partial_t u_\varepsilon + 3gu_\varepsilon^2 \partial_x u_\varepsilon - a\partial_x^3 u_\varepsilon + q\partial_x(v_\varepsilon u_\varepsilon) - \varepsilon\partial_x^5 u_\varepsilon \right) = bP_\varepsilon(t, \infty) = 0,$$

which gives (18).

Finally, we prove (19). Integrating the second equation of (16) on $(-\infty, x)$, by (16), we have that

$$P_\varepsilon(t, x) = \int_{-\infty}^{x} u_\varepsilon(t, y) dy. \tag{20}$$

Equation (19) follows from (18) and (20). □

Arguing as in ([15], Lemma 2.2), we have the following result.

**Lemma 2.** *Assume (5). We have that*

$$\int_{\mathbb{R}} u_\varepsilon^2 \partial_x v_\varepsilon dx = \begin{cases} \dfrac{\beta}{\kappa} \|\partial_x v_\varepsilon(t, \cdot)\|_{L^2(\mathbb{R})}^2, & \text{if } \beta \neq 0, \\ 0, & \text{if } \beta = 0. \end{cases} \tag{21}$$

**Lemma 3.** *Assume* (5). *If $\beta \neq 0$, then for each $t \geq 0$, there exists a constant $C_0 > 0$, independent on $\varepsilon$, such that*

$$\begin{aligned}\|u_\varepsilon(t,\cdot)\|^2_{L^2(\mathbb{R})} + \frac{q\beta}{\kappa}\int_0^t \|\partial_x v_\varepsilon(s,\cdot)\|^2_{L^2(\mathbb{R})} ds \\ + 2\varepsilon\int_0^t \|\partial_x u_\varepsilon(t,\cdot)\|^2_{L^2(\mathbb{R})} ds \leq C_0.\end{aligned} \tag{22}$$

*If $\beta = 0$, then for each $t \geq 0$,*

$$\|u_\varepsilon(t,\cdot)\|^2_{L^2(\mathbb{R})} + 2\varepsilon\int_0^t \|\partial_x u_\varepsilon(s,\cdot)\|^2_{L^2(\mathbb{R})} ds \leq \|u_0\|^2_{L^2(\mathbb{R})}. \tag{23}$$

*In particular, we have*

$$\begin{aligned}\|\partial_x v_\varepsilon(t,\cdot)\|_{L^\infty(\mathbb{R})}, \|\partial_x v_\varepsilon(t,\cdot)\|_{L^2(\mathbb{R})} \leq& C_0,\\ \|v_\varepsilon(t,\cdot)\|_{L^\infty(\mathbb{R})}, \|v_\varepsilon(t,\cdot)\|_{L^2(\mathbb{R})} \leq& C_0.\end{aligned} \tag{24}$$

*Moreover, fixed $T > 0$, there exists a constant $C(T) > 0$, independent on $\varepsilon$, such that*

$$\varepsilon\int_0^t \|\partial_x u_\varepsilon(s,\cdot)\|^2_{L^2(\mathbb{R})} ds \leq C(T). \tag{25}$$

**Proof.** Multiplying by $2u_\varepsilon$ the first equation of (16), an integration on $\mathbb{R}$ gives

$$\begin{aligned}\frac{d}{dt}\|u_\varepsilon(t,\cdot)\|^2_{L^2(\mathbb{R})} =& 2\int_{\mathbb{R}} u_\varepsilon \partial_t u_\varepsilon dx\\ =& -6g\int_{\mathbb{R}} u_\varepsilon^3\partial_x u_\varepsilon dx - 2q\int_{\mathbb{R}} u_\varepsilon\partial_x(u_\varepsilon v_\varepsilon)dx + 2b\int_{\mathbb{R}} P_\varepsilon u_\varepsilon dx\\ &+ 2a\int_{\mathbb{R}} u_\varepsilon\partial_x^3 u_\varepsilon dx - 2\varepsilon\int_{\mathbb{R}} u_\varepsilon\partial_x^4 u_\varepsilon dx\\ =& -2q\int_{\mathbb{R}} u_\varepsilon\partial_x(u_\varepsilon v_\varepsilon)dx - 2a\int_{\mathbb{R}}\partial_x u_\varepsilon\partial_x^2 u_\varepsilon dx\\ &+ 2b\int_{\mathbb{R}} P_\varepsilon u_\varepsilon dx + 2\varepsilon\int_{\mathbb{R}}\partial_x u_\varepsilon\partial_x^3 u_\varepsilon dx\\ =& -2q\int_{\mathbb{R}} u_\varepsilon\partial_x(u_\varepsilon v_\varepsilon)dx - 2a\int_{\mathbb{R}}\partial_x u_\varepsilon\partial_x^2 u_\varepsilon dx\\ &+ 2b\int_{\mathbb{R}} P_\varepsilon u_\varepsilon dx - 2\varepsilon\left\|\partial_x^2 u_\varepsilon(t,\cdot)\right\|^2_{L^2(\mathbb{R})}.\end{aligned}$$

Therefore,

$$\frac{d}{dt}\|u_\varepsilon(t,\cdot)\|^2_{L^2(\mathbb{R})} + 2\varepsilon\left\|\partial_x^2 u_\varepsilon(t,\cdot)\right\|^2_{L^2(\mathbb{R})} = 2b\int_{\mathbb{R}} P_\varepsilon u_\varepsilon dx - 2q\int_{\mathbb{R}} u_\varepsilon\partial_x(u_\varepsilon v_\varepsilon)dx.$$

Arguing as in ([15], Lemma 2.2), we have (22), (23) and (24).
Finally, arguing as in ([6], Lemma 2.3), we have (25). □

**Lemma 4.** *Assume* (5). *Fix $T > 0$. There exists a constant $C_0 > 0$, independent on $\varepsilon$, such that*

$$\left\|\partial_x^2 v_\varepsilon\right\|_{L^\infty((0,T)\times\mathbb{R})} \leq C_0\left(1 + \|u_\varepsilon\|^2_{L^\infty((0,T)\times\mathbb{R})}\right). \tag{26}$$

**Proof.** Let $0 \leq t \leq T$. Thanks to the third equation of (16), we have that

$$\alpha\partial_x^2 v_\varepsilon = \kappa u_\varepsilon^2 - \beta\partial_x v_\varepsilon - \gamma v_\varepsilon. \tag{27}$$

Therefore, by (24),

$$\begin{aligned}|\alpha||\partial_x^2 v_\varepsilon| =&|\kappa u_\varepsilon^2 - \beta\partial_x v_\varepsilon - \gamma v_\varepsilon| \le |\kappa|u_\varepsilon^2 + |\beta||\partial_x v_\varepsilon| + |\gamma||v_\varepsilon|\\ \le&|\kappa|\,\|u_\varepsilon\|^2_{L^2((0,T)\times\mathbb{R})} + |\beta|\,\|\partial_x v_\varepsilon(t,\cdot)\|_{L^\infty(\mathbb{R})} + |\gamma|\,\|v_\varepsilon(t,\cdot)\|_{L^\infty(\mathbb{R})}\\ \le&|\kappa|\,\|u_\varepsilon\|^2_{L^2((0,T)\times\mathbb{R})} + C_0 \le C_0\left(1+\|u_\varepsilon\|^2_{L^2((0,T)\times\mathbb{R})}\right),\end{aligned}$$

which gives (26). □

Arguing as in ([6], Lemma 2.2), we have the following result.

**Lemma 5.** *For each $t \ge 0$, we have that*

$$\int_0^{-\infty} P_\varepsilon(t,x)dx = A_\varepsilon(t), \tag{28}$$

$$\int_0^{\infty} P_\varepsilon(t,x)dx = A_\varepsilon(t), \tag{29}$$

*where*

$$A_\varepsilon(t) = -\frac{1}{b}\partial_t P_\varepsilon(t,0) - \frac{g}{b}u_\varepsilon^3(t,0) - \frac{a}{b}\partial_x^2 u_\varepsilon(t,0) - \frac{q}{b}u_\varepsilon(t,0)v_\varepsilon(t,0) + \frac{\varepsilon}{b}\partial_x u_\varepsilon(t,0).$$

*In particular, we have*

$$\int_{\mathbb{R}} P_\varepsilon(t,x)dx = 0. \tag{30}$$

**Lemma 6.** *Assume (5). Fix $T > 0$. There exists a constant $C(T) > 0$, independent on $\varepsilon$, such that*

$$\begin{aligned}\|P_\varepsilon(t,\cdot)\|^2_{L^2(\mathbb{R})} +&2\varepsilon e^{C_0 t}\int_0^t e^{-C_0 s}\,\|\partial_x u_\varepsilon(s,\cdot)\|^2_{L^2(\mathbb{R})}\,ds\\ \le& C(T)\left(1+\|u_\varepsilon\|^2_{L^\infty((0,T)\times\mathbb{R})}\right).\end{aligned} \tag{31}$$

*for every $0 \le t \le T$. In particular, we have that*

$$\|P_\varepsilon\|_{L^\infty((0,T)\times\mathbb{R})} \le \sqrt[4]{C(T)\left(1+\|u_\varepsilon\|^2_{L^\infty((0,T)\times\mathbb{R})}\right)}. \tag{32}$$

**Proof.** Let $0 \le t \le T$. We begin by observing that, by (28), we can consider the following function:

$$F_\varepsilon(t,x) = \int_{-\infty}^x P_\varepsilon(t,y)dy. \tag{33}$$

Integrating the second equation of (16) on $(-\infty, x)$, we have

$$P_\varepsilon(t,x) = \int_{-\infty}^x u_\varepsilon(t,y)dy. \tag{34}$$

Differentiating (34) with respect to $t$, we get

$$\partial_t P_\varepsilon(t,x) = \frac{d}{dt}\int_{-\infty}^x u_\varepsilon(t,y)dy = \int_{-\infty}^x \partial_t u_\varepsilon(t,x)dx. \tag{35}$$

Equation (33), (35) and an integration on $(-\infty, x)$ of the first equation of (16) give

$$\partial_t P_\varepsilon = bF_\varepsilon - \varepsilon\partial_x^3 u_\varepsilon - gu_\varepsilon^3 + a\partial_x^2 u_\varepsilon - qv_\varepsilon u_\varepsilon. \tag{36}$$

Multiplying (36) by $-2P_\varepsilon$, an integration on $\mathbb{R}$ gives

$$\begin{aligned}\frac{d}{dt}\|P_\varepsilon(t,\cdot)\|^2_{L^2(\mathbb{R})} =&2b\int_{\mathbb{R}} F_\varepsilon P_\varepsilon dx - 2\varepsilon\int_{\mathbb{R}} P_\varepsilon\partial_x^3 u_\varepsilon dx - 2g\int_{\mathbb{R}} P_\varepsilon u_\varepsilon^3 dx\\ &+ 2a\int_{\mathbb{R}} P_\varepsilon\partial_x^2 u_\varepsilon dx - 2q\int_{\mathbb{R}} P_\varepsilon v_\varepsilon u_\varepsilon dx.\end{aligned}\tag{37}$$

Observe that, by (18), (30), (33) and the second equation of (16),

$$\begin{aligned}2b\int_{\mathbb{R}} F_\varepsilon P_\varepsilon dx =&2b\int_{\mathbb{R}} F_\varepsilon\partial_x F_\varepsilon dx = bF_\varepsilon^2(t,\infty) = b\left(\int_{\mathbb{R}} P_\varepsilon(t,x)dx\right)^2 dx = 0,\\ -2\varepsilon\int_{\mathbb{R}} P_\varepsilon\partial_x^3 u_\varepsilon dx =&2\varepsilon\int_{\mathbb{R}}\partial_x P_\varepsilon\partial_x^2 u_\varepsilon dx = 2\varepsilon\int_{\mathbb{R}} u_\varepsilon\partial_x^2 u_\varepsilon dx = -2\varepsilon\|\partial_x u_\varepsilon(t,\cdot)\|^2_{L^2(\mathbb{R})},\\ 2a\int_{\mathbb{R}} P_\varepsilon\partial_x^2 u_\varepsilon dx =&-2a\int_{\mathbb{R}}\partial_x P_\varepsilon\partial_x u_\varepsilon dx = -2a\int_{\mathbb{R}} u_\varepsilon\partial_x u_\varepsilon = 0,\\ -2q\int_{\mathbb{R}} P_\varepsilon v_\varepsilon u_\varepsilon dx =&-2q\int_{\mathbb{R}} P_\varepsilon v_\varepsilon\partial_x P_\varepsilon dx = 2q\int_{\mathbb{R}}\partial_x v_\varepsilon P_\varepsilon^2 dx.\end{aligned}$$

Consequently, by (24) and (37),

$$\begin{aligned}&\frac{d}{dt}\|P_\varepsilon(t,\cdot)\|^2_{L^2(\mathbb{R})} + 2\varepsilon\|\partial_x u_\varepsilon(t,\cdot)\|^2_{L^2(\mathbb{R})}\\ &\quad=2q\int_{\mathbb{R}}\partial_x v_\varepsilon P_\varepsilon^2 dx - 2g\int_{\mathbb{R}} P_\varepsilon u_\varepsilon^3 dx\\ &\quad\le 2|q|\int_{\mathbb{R}}|\partial_x v_\varepsilon|P_\varepsilon^2 dx + 2|g|\int_{\mathbb{R}}|P_\varepsilon||u_\varepsilon|^3 dx\\ &\quad\le 2|q|\|\partial_x v_\varepsilon(t,\cdot)\|_{L^\infty(\mathbb{R})}\|P_\varepsilon(t,\cdot)\|^2_{L^2(\mathbb{R})} + 2|g|\int_{\mathbb{R}}|P_\varepsilon||u_\varepsilon|^3 dx\\ &\quad\le C_0\|P_\varepsilon(t,\cdot)\|^2_{L^2(\mathbb{R})} + 2|g|\int_{\mathbb{R}}|P_\varepsilon||u_\varepsilon|^3 dx.\end{aligned}\tag{38}$$

Due to Lemma 3 and the Young inequality,

$$\begin{aligned}2|g|\int_{\mathbb{R}}|P_\varepsilon||u_\varepsilon|^3 dx =&\int_{\mathbb{R}}|2P_\varepsilon u_\varepsilon||gu_\varepsilon^2|dx \le 2\int_{\mathbb{R}} P_\varepsilon^2 u_\varepsilon^2 dx + \frac{g^2}{2}\int_{\mathbb{R}} u_\varepsilon^4 dx\\ \le&2\|P_\varepsilon(t,\cdot)\|^2_{L^\infty(\mathbb{R})}\|u_\varepsilon(t,\cdot)\|^2_{L^2(\mathbb{R})}\\ &+\frac{g^2}{2}\|u_\varepsilon\|^2_{L^\infty((0,T)\times\mathbb{R})}\|u_\varepsilon(t,\cdot)\|^2_{L^2(\mathbb{R})}\\ \le&C_0\|P_\varepsilon(t,\cdot)\|^4_{L^\infty(\mathbb{R})} + C_0\|u_\varepsilon\|^2_{L^\infty((0,T)\times\mathbb{R})}\\ \le&\|P_\varepsilon(t,\cdot)\|^4_{L^\infty(\mathbb{R})} + C_0 + C_0\|u_\varepsilon\|^2_{L^\infty((0,T)\times\mathbb{R})}.\end{aligned}$$

It follows from (38) that

$$\begin{aligned}&\frac{d}{dt}\|P_\varepsilon(t,\cdot)\|^2_{L^2(\mathbb{R})} + 2\varepsilon\|\partial_x u_\varepsilon(t,\cdot)\|^2_{L^2(\mathbb{R})}\\ &\quad\le C_0\|P_\varepsilon(t,\cdot)\|^2_{L^2(\mathbb{R})} + \|P_\varepsilon(t,\cdot)\|^4_{L^\infty(\mathbb{R})} + C_0\left(1+\|u_\varepsilon\|^2_{L^\infty((0,T)\times\mathbb{R})}\right).\end{aligned}\tag{39}$$

Thanks to Lemma 3 and the Hölder inequality,

$$\begin{aligned}P_\varepsilon^2(t,x) =& 2\int_{-\infty}^x P_\varepsilon u_\varepsilon dy \le 2\int_{\mathbb{R}}|P_\varepsilon||u_\varepsilon|dx\\ \le&\|P_\varepsilon(t,\cdot)\|_{L^2(\mathbb{R})}\|u_\varepsilon(t,\cdot)\|_{L^2(\mathbb{R})} \le C_0\|P_\varepsilon(t,\cdot)\|_{L^2(\mathbb{R})}.\end{aligned}$$

Hence,

$$\|P_\varepsilon(t,\cdot)\|^4_{L^\infty(\mathbb{R})} \le C_0 \|P_\varepsilon(t,\cdot)\|^2_{L^2(\mathbb{R})}. \tag{40}$$

It follows from (39) and (40) that

$$\begin{aligned}\frac{d}{dt}\|P_\varepsilon(t,\cdot)\|^2_{L^2(\mathbb{R})} &+ 2\varepsilon\|u_\varepsilon(t,\cdot)\|^2_{L^2(\mathbb{R})}\\ &\le C_0\|P_\varepsilon(t,\cdot)\|^2_{L^2(\mathbb{R})} + C_0\left(1+\|u_\varepsilon\|^2_{L^\infty((0,T)\times\mathbb{R})}\right).\end{aligned}$$

The Gronwall Lemma and (17) give

$$\begin{aligned}\|P_\varepsilon(t,\cdot)\|^2_{L^2(\mathbb{R})} + 2\varepsilon e^{C_0 t}\int_0^t e^{-C_0 s}\|u_\varepsilon(s,\cdot)\|^2_{L^2(\mathbb{R})}ds&\\ \le & C_0 e^{C_0 t} + C_0\left(1+\|u_\varepsilon\|^2_{L^\infty((0,T)\times\mathbb{R})}\right)e^{C_0 t}\int_0^t e^{-C_0 s}ds\\ \le & C(T)\left(1+\|u_\varepsilon\|^2_{L^\infty((0,T)\times\mathbb{R})}\right),\end{aligned}$$

which gives (31).

Finally, (32) follows from (31) and (40). □

Following ([6], Lemma 2.5), we prove the following result.

**Lemma 7.** *Assume (5). Fix $T > 0$. There exists a constant $C(T) > 0$, independent on $\varepsilon$, such that*

$$\|u_\varepsilon\|_{L^\infty((0,T)\times\mathbb{R})} \le C(T). \tag{41}$$

*In particular, we have*

$$\|\partial_x u_\varepsilon(t,\cdot)\|^2_{L^2(\mathbb{R})} + 2\varepsilon e^{C_0 t}\int_0^t e^{-C_0 s}\left\|\partial_x^3 u_\varepsilon(s,\cdot)\right\|^2_{L^2(\mathbb{R})}ds \le C(T), \tag{42}$$

*for every $0 \le t \le T$. Moreover,*

$$\begin{aligned}\left\|\partial_x^2 v_\varepsilon\right\|_{L^\infty((0,T)\times\mathbb{R})} &\le C(T),\\ \|P_\varepsilon(t,\cdot)\|_{L^2(\mathbb{R})} &\le C(T),\\ \|P_\varepsilon\|_{L^\infty((0,T)\times\mathbb{R})} &\le C(T),\end{aligned} \tag{43}$$

*for every $0 \le t \le T$.*

**Proof.** Let $0 \le t \le T$. Multiplying the first equation of (1) by $-2\partial_x^2 u_\varepsilon + \frac{2g}{a}u_\varepsilon^3$, we have that

$$\begin{aligned}\left(-2\partial_x^2 u_\varepsilon + \frac{2g}{a}u_\varepsilon^3\right)\partial_t u_\varepsilon &+ 3g\left(-2\partial_x^2 u_\varepsilon + \frac{2g}{a}u_\varepsilon^3\right)u_\varepsilon^2\partial_x u_\varepsilon\\ &- a\left(-2\partial_x^2 u_\varepsilon + \frac{2g}{a}u_\varepsilon^3\right)\partial_x^3 u_\varepsilon + q\left(-2\partial_x^2 u_\varepsilon + \frac{2g}{a}u_\varepsilon^3\right)\partial_x(v_\varepsilon u_\varepsilon)\\ =& b\left(-2\partial_x^2 u_\varepsilon + \frac{2g}{a}u_\varepsilon^3\right)P_\varepsilon - \varepsilon\left(-2\partial_x^2 u_\varepsilon + \frac{2g}{a}u_\varepsilon^3\right)\partial_x^4 u_\varepsilon.\end{aligned} \tag{44}$$

Observe that, by (18) and the second equation of (16),

$$-2b\int_{\mathbb{R}} P_\varepsilon\partial_x^2 u_\varepsilon dx = 2b\int_{\mathbb{R}}\partial_x P_\varepsilon\partial_x u_\varepsilon dx = 2b\int_{\mathbb{R}} u_\varepsilon\partial_x u_\varepsilon dx = 0. \tag{45}$$

Moreover,

$$
\begin{aligned}
\int_{\mathbb{R}}\left(-2\partial_x^2 u_\varepsilon+\frac{2g}{a}u_\varepsilon^3\right)\partial_t u_\varepsilon dx =&\frac{d}{dt}\left(\|\partial_x u_\varepsilon(t,\cdot)\|^2_{L^2(\mathbb{R})}+\frac{g}{2a}\int_{\mathbb{R}}u_\varepsilon^4 dx\right),\\
3g\int_{\mathbb{R}}\left(-2\partial_x^2 u_\varepsilon+\frac{2g}{a}u_\varepsilon^3\right)u_\varepsilon^2\partial_x u_\varepsilon dx =&-6g\int_{\mathbb{R}}u_\varepsilon^2\partial_x u_\varepsilon\partial_x^2 u_\varepsilon dx,\\
-a\int_{\mathbb{R}}\left(-2\partial_x^2 u_\varepsilon+\frac{2g}{a}u_\varepsilon^3\right)\partial_x^3 u_\varepsilon dx =&6g\int_{\mathbb{R}}u_\varepsilon^2\partial_x u_\varepsilon\partial_x^2 u_\varepsilon dx,\\
-\varepsilon\int_{\mathbb{R}}\left(-2\partial_x^2 u_\varepsilon+\frac{2g}{a}u_\varepsilon^3\right)\partial_x^4 u_\varepsilon dx =&-2\varepsilon\left\|\partial_x^3 u_\varepsilon(t,\cdot)\right\|^2_{L^2(\mathbb{R})}\\
&+\frac{6g\varepsilon}{a}\int_{\mathbb{R}}u_\varepsilon^2\partial_x u_\varepsilon\partial_x^3 u_\varepsilon dx.
\end{aligned}
\tag{46}
$$

Defined

$$
G(t):=\|\partial_x u_\varepsilon(t,\cdot)\|^2_{L^2(\mathbb{R})}+\frac{g}{2a}\int_{\mathbb{R}}u_\varepsilon^4 dx,
\tag{47}
$$

it follows from (45), (46) and an integration on $\mathbb{R}$ of (44) that

$$
\begin{aligned}
\frac{dG(t)}{dt}+2\varepsilon\left\|\partial_x^3 u_\varepsilon(t,\cdot)\right\|^2_{L^2(\mathbb{R})}=&\frac{2bg}{a}\int_{\mathbb{R}}P_\varepsilon u_\varepsilon^3 dx+\frac{6g\varepsilon}{a}\int_{\mathbb{R}}u_\varepsilon^2\partial_x u_\varepsilon\partial_x^3 u_\varepsilon dx\\
&+2q\int_{\mathbb{R}}\partial_x(v_\varepsilon u_\varepsilon)\partial_x^2 u_\varepsilon dx-2qga\int_{\mathbb{R}}\partial_x(v_\varepsilon u_\varepsilon)u_\varepsilon^3 dx.
\end{aligned}
\tag{48}
$$

Observe that

$$
\begin{aligned}
2q\int_{\mathbb{R}}\partial_x(v_\varepsilon u_\varepsilon)\partial_x^2 u_\varepsilon dx =&2q\int_{\mathbb{R}}u_\varepsilon\partial_x v_\varepsilon\partial_x^2 u_\varepsilon dx+2q\int_{\mathbb{R}}v_\varepsilon\partial_x u_\varepsilon\partial_x^2 u_\varepsilon dx\\
=&-2q\int_{\mathbb{R}}\partial_x^2 v_\varepsilon u_\varepsilon\partial_x u_\varepsilon dx-3q\int_{\mathbb{R}}\partial_x v_\varepsilon(\partial_x u_\varepsilon)^2 dx,\\
-2qga\int_{\mathbb{R}}\partial_x(v_\varepsilon u_\varepsilon)u_\varepsilon^3 dx =&-2qga\int_{\mathbb{R}}\partial_x v_\varepsilon u_\varepsilon^4 dx-2qga\int_{\mathbb{R}}v_\varepsilon\partial_x u_\varepsilon u_\varepsilon^3 dx\\
&-\frac{3qg}{2a}\int_{\mathbb{R}}\partial_x v_\varepsilon u_\varepsilon^4 dx.
\end{aligned}
$$

Consequently, by (48),

$$
\begin{aligned}
\frac{dG(t)}{dt}+2\varepsilon\left\|\partial_x^3 u_\varepsilon(t,\cdot)\right\|^2_{L^2(\mathbb{R})}=&\frac{2bg}{a}\int_{\mathbb{R}}P_\varepsilon u_\varepsilon^3 dx+\frac{6g\varepsilon}{a}\int_{\mathbb{R}}u_\varepsilon^2\partial_x u_\varepsilon\partial_x^3 u_\varepsilon dx\\
&-3q\int_{\mathbb{R}}\partial_x v_\varepsilon(\partial_x u_\varepsilon)^2 dx-2q\int_{\mathbb{R}}\partial_x^2 v_\varepsilon u_\varepsilon\partial_x u_\varepsilon dx\\
&-\frac{3qg}{2a}\int_{\mathbb{R}}\partial_x v_\varepsilon u_\varepsilon^4 dx.
\end{aligned}
\tag{49}
$$

Due to (26), (32), Lemma 3 and the Young inequality,

$$
\begin{aligned}
\left|\frac{2bg}{a}\right|\int_{\mathbb{R}} P_\varepsilon u_\varepsilon^3 dx =&\int_{\mathbb{R}} |2P_\varepsilon u_\varepsilon| \left|\frac{bgu_\varepsilon^2}{a}\right| dx\\
\leq& 2\int_{\mathbb{R}} P_\varepsilon^2 u_\varepsilon^2 dx + \frac{b^2g^2}{2a^2}\int_{\mathbb{R}} u_\varepsilon^4 dx\\
\leq& 2\left\|P_\varepsilon\right\|^2_{L^\infty((0,T)\times\mathbb{R})}\left\|u_\varepsilon(t,\cdot)\right\|^2_{L^2(\mathbb{R})}\\
&+\frac{b^2g^2}{2a^2}\left\|u_\varepsilon\right\|^2_{L^\infty((0,T)\times\mathbb{R})}\left\|u_\varepsilon(t,\cdot)\right\|^2_{L^2(\mathbb{R})}\\
\leq& 2C_0\left\|P_\varepsilon\right\|^2_{L^\infty((0,T)\times\mathbb{R})} + C_0\left\|u_\varepsilon\right\|^2_{L^\infty((0,T)\times\mathbb{R})}\\
\leq& \left\|P_\varepsilon\right\|^4_{L^\infty((0,T)\times\mathbb{R})} + C_0\left\|u_\varepsilon\right\|^2_{L^\infty((0,T)\times\mathbb{R})} + C_0\\
\leq& C(T)\left(1+\left\|u_\varepsilon\right\|^2_{L^\infty((0,T)\times\mathbb{R})}\right),\\
3|q|\int_{\mathbb{R}} |\partial_x v_\varepsilon|(\partial_x u_\varepsilon)^2 dx \leq& 3|q|\left\|\partial_x v_\varepsilon(t,\cdot)\right\|_{L^\infty(\mathbb{R})}\left\|\partial_x u_\varepsilon(t,\cdot)\right\|^2_{L^2(\mathbb{R})}\\
\leq& C_0\left\|\partial_x u_\varepsilon(t,\cdot)\right\|^2_{L^2(\mathbb{R})},\\
2|q|\int_{\mathbb{R}} \partial_x^2 v_\varepsilon u_\varepsilon \partial_x u_\varepsilon dx =& 2\int_{\mathbb{R}} |q\partial_x^2 v_\varepsilon u_\varepsilon| \partial_x u_\varepsilon| dx\\
\leq& q^2\int_{\mathbb{R}} (\partial_x^2 v_\varepsilon)^2 u_\varepsilon^2 dx + \left\|\partial_x u_\varepsilon(t,\cdot)\right\|^2_{L^2(\mathbb{R})}\\
\leq& q^2\left\|\partial_x^2 v_\varepsilon\right\|^2_{L^\infty((0,T)\times\mathbb{R})}\left\|u_\varepsilon(t,\cdot)\right\|^2_{L^2(\mathbb{R})} + \left\|\partial_x u_\varepsilon(t,\cdot)\right\|^2_{L^2(\mathbb{R})}\\
\leq& C_0\left\|\partial_x^2 v_\varepsilon\right\|^2_{L^\infty((0,T)\times\mathbb{R})} + \left\|\partial_x u_\varepsilon(t,\cdot)\right\|^2_{L^2(\mathbb{R})}\\
\leq& C(T)\left(1+\left\|u_\varepsilon\right\|^2_{L^\infty((0,T)\times\mathbb{R})}\right) + \left\|\partial_x u_\varepsilon(t,\cdot)\right\|^2_{L^2(\mathbb{R})},\\
\left|\frac{3qg}{2a}\right|\int_{\mathbb{R}} |\partial_x v_\varepsilon| u_\varepsilon^4 dx \leq& \left|\frac{3qg}{2a}\right|\left\|\partial_x v_\varepsilon(t,\cdot)\right\|_{L^\infty(\mathbb{R})}\int_{\mathbb{R}} u_\varepsilon^4 dx\\
\leq& C_0\int_{\mathbb{R}} u_\varepsilon^4 dx \leq C_0\left\|u_\varepsilon^2\right\|_{L^\infty((0,T)\times\mathbb{R})}\left\|u_\varepsilon(t,\cdot)\right\|^2_{L^2(\mathbb{R})}\\
\leq& C_0\left\|u_\varepsilon^2\right\|_{L^\infty((0,T)\times\mathbb{R})},\\
\left|\frac{6g\varepsilon}{a}\right|\int_{\mathbb{R}} |u_\varepsilon^2\partial_x u_\varepsilon||\partial_x^3 u_\varepsilon| dx =& 2\varepsilon\int_{\mathbb{R}}\left|\frac{3g}{a}u_\varepsilon^2\partial_x u_\varepsilon\right|\left|\partial_x^3 u_\varepsilon\right| dx\\
\leq& \frac{9g^2\varepsilon}{a^2}\int_{\mathbb{R}} u_\varepsilon^4(\partial_x u_\varepsilon)^2 dx + \varepsilon\left\|\partial_x^3 u_\varepsilon(t,\cdot)\right\|^2_{L^2(\mathbb{R})}.
\end{aligned}
$$

Consequently, by (49),

$$
\begin{aligned}
\frac{dG(t)}{dt} + \varepsilon\left\|\partial_x^3 u_\varepsilon(t,\cdot)\right\|^2_{L^2(\mathbb{R})} \leq& C_0\left\|\partial_x u_\varepsilon(t,\cdot)\right\|^2_{L^2(\mathbb{R})} + \frac{9g^2\varepsilon}{a^2}\int_{\mathbb{R}} u_\varepsilon^4(\partial_x u_\varepsilon)^2 dx\\
&+ C(T)\left(1+\left\|u_\varepsilon\right\|^2_{L^\infty((0,T)\times\mathbb{R})}\right).
\end{aligned}
\tag{50}
$$

Lemma 2.6 of [6] says that

$$
\int_{\mathbb{R}} u_\varepsilon^4(\partial_x u_\varepsilon)^2 dx \leq 4\left\|u_\varepsilon(t,\cdot)\right\|^4_{L^2(\mathbb{R})}\left\|\partial_x^2 u_\varepsilon(t,\cdot)\right\|^2_{L^2(\mathbb{R})}. \tag{51}
$$

Hence, by Lemma 3, we have that

$$\begin{aligned}\frac{9g^2\varepsilon}{a^2}\int_{\mathbb{R}} u_\varepsilon^4(\partial_x u_\varepsilon)^2dx \leq& \frac{36g^2\varepsilon}{a^2}\left\|u_\varepsilon(t,\cdot)\right\|^4_{L^2(\mathbb{R})}\left\|\partial_x^2 u_\varepsilon(t,\cdot)\right\|^2_{L^2(\mathbb{R})}\\ \leq& C_0\varepsilon\left\|\partial_x^2 u_\varepsilon(t,\cdot)\right\|^2_{L^2(\mathbb{R})}.\end{aligned}$$

Therefore, by (50),

$$\begin{aligned}\frac{dG(t)}{dt}+\varepsilon\left\|\partial_x^3 u_\varepsilon(t,\cdot)\right\|^2_{L^2(\mathbb{R})} \leq& C_0\left\|\partial_x u_\varepsilon(t,\cdot)\right\|^2_{L^2(\mathbb{R})}+C_0\varepsilon\left\|\partial_x^2 u_\varepsilon(t,\cdot)\right\|^2_{L^2(\mathbb{R})}\\ &+C(T)\left(1+\left\|u_\varepsilon\right\|^2_{L^\infty((0,T)\times\mathbb{R})}\right).\end{aligned} \tag{52}$$

Observe that, by (47) and Lemma 3,

$$\begin{aligned}C_0\left\|\partial_x u_\varepsilon(t,\cdot)\right\|^2_{L^2(\mathbb{R})} =& C_0G(t)-\frac{gC_0}{2a}\int_{\mathbb{R}} u_\varepsilon^4dx\\ \leq& C_0G(t)+\left|\frac{gC_0}{2a}\right|\left\|u_\varepsilon\right\|^2_{L^\infty((0,T)\times\mathbb{R})}\left\|u_\varepsilon(t,\cdot)\right\|^2_{L^2(\mathbb{R})}\\ \leq& C_0G(t)+C_0\left\|u_\varepsilon\right\|^2_{L^\infty((0,T)\times\mathbb{R})}.\end{aligned} \tag{53}$$

It follows from (52) and (53) that

$$\begin{aligned}\frac{dG(t)}{dt}+\varepsilon\left\|\partial_x^3 u_\varepsilon(t,\cdot)\right\|^2_{L^2(\mathbb{R})} \leq& C_0G(t)+C_0\varepsilon\left\|\partial_x^2 u_\varepsilon(t,\cdot)\right\|^2_{L^2(\mathbb{R})}\\ &+C(T)\left(1+\left\|u_\varepsilon\right\|^2_{L^\infty((0,T)\times\mathbb{R})}\right).\end{aligned}$$

The Gronwall Lemma, (17), (47) and Lemma 3 that

$$\begin{aligned}&\left\|\partial_x u_\varepsilon(t,\cdot)\right\|^2_{L^2(\mathbb{R})}+\frac{g}{2a}\int_{\mathbb{R}} u_\varepsilon^4dx+\varepsilon e^{C_0t}\int_0^t e^{-C_0s}\left\|\partial_x^3 u_\varepsilon(s,\cdot)\right\|^2_{L^2(\mathbb{R})}ds\\ &\leq C_0e^{C_0t}+C(T)\left(1+\left\|u_\varepsilon\right\|^2_{L^\infty((0,T)\times\mathbb{R})}\right)e^{C_0t}\int_0^t e^{-C_0s}ds\\ &\quad+C_0\varepsilon e^{C_0t}\int_0^t e^{-C_0s}\left\|\partial_x^2 u_\varepsilon(s,\cdot)\right\|^2_{L^2(\mathbb{R})}ds\\ &\leq C(T)\left(1+\left\|u_\varepsilon\right\|^2_{L^\infty((0,T)\times\mathbb{R})}\right)+C(T)\varepsilon\int_0^t\left\|\partial_x^2 u_\varepsilon(s,\cdot)\right\|^2_{L^2(\mathbb{R})}ds\\ &\leq C(T)\left(1+\left\|u_\varepsilon\right\|^2_{L^\infty((0,T)\times\mathbb{R})}\right).\end{aligned}$$

Consequently, by Lemma 3,

$$\begin{aligned}&\left\|\partial_x u_\varepsilon(t,\cdot)\right\|^2_{L^2(\mathbb{R})}+2\varepsilon e^{C_0t}\int_0^t e^{-C_0s}\left\|\partial_x^3 u_\varepsilon(s,\cdot)\right\|^2_{L^2(\mathbb{R})}ds\\ &\quad\leq C(T)\left(1+\left\|u_\varepsilon\right\|^2_{L^\infty((0,T)\times\mathbb{R})}\right)-\frac{g}{2a}\left\|u_\varepsilon(t,\cdot)\right\|^4_{L^4(\mathbb{R})}\\ &\quad\leq C(T)\left(1+\left\|u_\varepsilon\right\|^2_{L^\infty((0,T)\times\mathbb{R})}\right)+\left|\frac{g}{2a}\right|\int_{\mathbb{R}} u_\varepsilon^4dx\\ &\quad\leq C(T)\left(1+\left\|u_\varepsilon\right\|^2_{L^\infty((0,T)\times\mathbb{R})}\right)+\left|\frac{g}{2a}\right|\left\|u_\varepsilon\right\|^2_{L^\infty((0,T)\times\mathbb{R})}\left\|u_\varepsilon(t,\cdot)\right\|^2_{L^2(\mathbb{R})}\\ &\quad\leq C(T)\left(1+\left\|u_\varepsilon\right\|^2_{L^\infty((0,T)\times\mathbb{R})}\right)+C_0\left\|u_\varepsilon\right\|^2_{L^\infty((0,T)\times\mathbb{R})}\\ &\quad\leq C(T)\left(1+\left\|u_\varepsilon\right\|^2_{L^\infty((0,T)\times\mathbb{R})}\right).\end{aligned} \tag{54}$$

We prove (41). Thanks to (54), Lemma 3 and the Hölder inequality,

$$\begin{aligned}u_\varepsilon^2(t,x) =&2\int_{-\infty}^{x} u_\varepsilon\partial_x u_\varepsilon dx \leq \int_{\mathbb{R}} |u_\varepsilon||\partial_x u_\varepsilon| dx\\ &\leq \|u_\varepsilon(t,\cdot)\|^2_{L^2(\mathbb{R})}\|\partial_x u_\varepsilon(t,\cdot)\|^2_{L^2(\mathbb{R})} \leq C(T)\sqrt{\left(1+\|u_\varepsilon\|^2_{L^\infty((0,T)\times\mathbb{R})}\right)}.\end{aligned}$$

Hence,

$$\|u_\varepsilon\|^4_{L^\infty((0,T)\times\mathbb{R})} - C(T)\|u_\varepsilon\|^2_{L^\infty((0,T)\times\mathbb{R})} - C(T) \leq 0,$$

which gives (41).

Finally, (42) follows from (41) and (54), while (26), (31), (32) and (41) give (43). □

Arguing as in ([15], Lemmas 2.8 and 2.9), we have the following result.

**Lemma 8.** *Assume (5). Fix $T > 0$. There exists a constant $C(T) > 0$, independent on $\varepsilon$, such that*

$$\left\|\partial_x^2 v_\varepsilon(t,\cdot)\right\|_{L^2(\mathbb{R})}, \left\|\partial_x^3 v_\varepsilon(t,\cdot)\right\|_{L^2(\mathbb{R})} \leq C(T), \tag{55}$$

*for every $0 \leq t \leq T$.*

**Lemma 9.** *Assume (5). Fix $T > 0$. There exists a constant $C(T) > 0$, independent on $\varepsilon$, such that*

$$\|\partial_x u_\varepsilon\|_{L^\infty((0,T)\times\mathbb{R})} \leq C(T), \tag{56}$$

*In particular, we have that*

$$\left\|\partial_x^2 u_\varepsilon(t,\cdot)\right\|^2_{L^2(\mathbb{R})} + 2\varepsilon e^{C_0 t}\int_0^t e^{-C_0 s}\left\|\partial_x^4 u_\varepsilon(s,\cdot)\right\|^2_{L^2(\mathbb{R})} ds \leq C(T), \tag{57}$$

*for every $0 \leq t \leq T$.*

**Proof.** Let $0 \leq t \leq T$. Consider two real constants $D$, $E$ which will be specified later Multiplying the first equation of (16) by

$$2a^2\partial_x^4 u_\varepsilon + Dagu_\varepsilon(\partial_x u_\varepsilon)^2 + Eagu_\varepsilon^2\partial_x^2 u_\varepsilon,$$

we have that

$$\begin{aligned}&\left(2a^2\partial_x^4 u_\varepsilon + Dagu_\varepsilon(\partial_x u_\varepsilon)^2 + Eagu_\varepsilon^2\partial_x^2 u_\varepsilon\right)\partial_t u_\varepsilon\\ &\quad + 3g\left(2a^2\partial_x^4 u_\varepsilon + Dagu_\varepsilon(\partial_x u_\varepsilon)^2 + Eagu_\varepsilon^2\partial_x^2 u_\varepsilon\right)u_\varepsilon^2\partial_x u_\varepsilon\\ &\quad - a\left(2a^2\partial_x^4 u_\varepsilon + Dagu_\varepsilon(\partial_x u_\varepsilon)^2 + Eagu_\varepsilon^2\partial_x^2 u_\varepsilon\right)\partial_x^3 u_\varepsilon\\ &\quad + q\left(2a^2\partial_x^4 u_\varepsilon + Dagu_\varepsilon(\partial_x u_\varepsilon)^2 + Eagu_\varepsilon^2\partial_x^2 u_\varepsilon\right)u_\varepsilon\partial_x v_\varepsilon\\ &\quad + q\left(2a^2\partial_x^4 u_\varepsilon + Dagu_\varepsilon(\partial_x u_\varepsilon)^2 + Eagu_\varepsilon^2\partial_x^2 u_\varepsilon\right)v_\varepsilon\partial_x u_\varepsilon\\ &\quad = b\left(2a^2\partial_x^4 u_\varepsilon + Dagu_\varepsilon(\partial_x u_\varepsilon)^2 + Eagu_\varepsilon^2\partial_x^2 u_\varepsilon\right)P_\varepsilon\\ &\quad - \varepsilon\left(2a^2\partial_x^4 u_\varepsilon + Dagu_\varepsilon(\partial_x u_\varepsilon)^2 + Eagu_\varepsilon^2\partial_x^2 u_\varepsilon\right)\partial_x^4 u_\varepsilon.\end{aligned} \tag{58}$$

Observe that

$$\begin{aligned}
&\int_{\mathbb{R}}\left(2a^2\partial_x^4 u_\varepsilon + Dagu_\varepsilon(\partial_x u_\varepsilon)^2 + Eagu_\varepsilon^2\partial_x^2 u_\varepsilon\right)\partial_t u_\varepsilon dx\\
&\quad = a^2\frac{d}{dt}\left\|\partial_x^2 u_\varepsilon(t,\cdot)\right\|^2_{L^2(\mathbb{R})} + Dag\int_{\mathbb{R}} u_\varepsilon(\partial_x u_\varepsilon)^2\partial_t u_\varepsilon dx + Eag\int_{\mathbb{R}} u_\varepsilon^2\partial_x^2 u_\varepsilon\partial_t u_\varepsilon dx,\\
&3g\int_{\mathbb{R}}\left(2a^2\partial_x^4 u_\varepsilon + Dagu_\varepsilon(\partial_x u_\varepsilon)^2 + Eagu_\varepsilon^2\partial_x^2 u_\varepsilon\right)u_\varepsilon^2\partial_x u_\varepsilon dx\\
&\quad = -12a^2 g\int_{\mathbb{R}} u_\varepsilon(\partial_x u_\varepsilon)^2\partial_x^3 u_\varepsilon dx - 6a^2 g\int_{\mathbb{R}} u_\varepsilon^2\partial_x^2 u_\varepsilon\partial_x^3 u_\varepsilon dx\\
&\qquad + (3D-6E)\,ag^2\int_{\mathbb{R}} u_\varepsilon^3(\partial_x u_\varepsilon)^3 dx\\
&\quad = 30a^2 g\int_{\mathbb{R}} u_\varepsilon\partial_x u_\varepsilon(\partial_x^2 u_\varepsilon)^2 dx + (3D-6E)\,ag^2\int_{\mathbb{R}} u_\varepsilon^3(\partial_x u_\varepsilon)^3 dx,\\
&-a\int_{\mathbb{R}}\left(2a^2\partial_x^4 u_\varepsilon + Dagu_\varepsilon(\partial_x u_\varepsilon)^2 + Eagu_\varepsilon^2\partial_x^2 u_\varepsilon\right)\partial_x^3 u_\varepsilon dx\\
&\quad = -(2D+E)\,a^2 g\int_{\mathbb{R}} u_\varepsilon\partial_x u_\varepsilon(\partial_x^2 u_\varepsilon)^2 dx,\\
&q\int_{\mathbb{R}}\left(2a^2\partial_x^4 u_\varepsilon + Dagu_\varepsilon(\partial_x u_\varepsilon)^2 + Eagu_\varepsilon^2\partial_x^2 u_\varepsilon\right)u_\varepsilon\partial_x v_\varepsilon dx\\
&\quad = -2a^2 q\int_{\mathbb{R}}\partial_x u_\varepsilon\partial_x v_\varepsilon\partial_x^3 u_\varepsilon dx - 2a^2 q\int_{\mathbb{R}} u_\varepsilon\partial_x^2 v_\varepsilon\partial_x^3 u_\varepsilon dx\\
&\qquad + (D-3E)\,agq\int_{\mathbb{R}} u_\varepsilon^2(\partial_x u_\varepsilon)^2\partial_x v_\varepsilon dx - agqE\int_{\mathbb{R}} u_\varepsilon^3\partial_x u_\varepsilon\partial_x^2 v_\varepsilon dx\\
&\quad = 2a^2 q\int_{\mathbb{R}}\partial_x v_\varepsilon(\partial_x^2 u_\varepsilon)^2 dx + 4a^2 q\int_{\mathbb{R}}\partial_x u_\varepsilon\partial_x^2 v_\varepsilon\partial_x^2 u_\varepsilon dx\\
&\qquad + 2a^2 q\int_{\mathbb{R}} u_\varepsilon\partial_x^3 v_\varepsilon\partial_x^2 u_\varepsilon dx + aq\,(D-3E)\int_{\mathbb{R}} u_\varepsilon^2(\partial_x u_\varepsilon)^2\partial_x v_\varepsilon dx,\\
&q\int_{\mathbb{R}}\left(2a^2\partial_x^4 u_\varepsilon + Dagu_\varepsilon(\partial_x u_\varepsilon)^2 + Eagu_\varepsilon^2\partial_x^2 u_\varepsilon\right)v_\varepsilon\partial_x u_\varepsilon dx\\
&\quad = -2a^2 q\int_{\mathbb{R}}\partial_x v_\varepsilon\partial_x u_\varepsilon\partial_x^3 u_\varepsilon dx - 2a^2 q\int_{\mathbb{R}} v_\varepsilon\partial_x^2 u_\varepsilon\partial_x^3 u_\varepsilon dx\\
&\qquad + (D-E)\,agq\int_{\mathbb{R}} u_\varepsilon v_\varepsilon(\partial_x u_\varepsilon)^3 dx - \frac{Eagq}{2}\int_{\mathbb{R}} u_\varepsilon^2(\partial_x u_\varepsilon)^3\partial_x^2 v_\varepsilon dx\\
&\quad = 2a^2 q\int_{\mathbb{R}}\partial_x^2 v_\varepsilon\partial_x u_\varepsilon\partial_x^2 u_\varepsilon dx + 3a^2 q\int_{\mathbb{R}}\partial_x v_\varepsilon(\partial_x^2 u_\varepsilon)^2 dx\\
&\qquad + (D-E)\,agq\int_{\mathbb{R}} u_\varepsilon v_\varepsilon(\partial_x u_\varepsilon)^3 dx - \frac{Eagq}{2}\int_{\mathbb{R}} u_\varepsilon^2(\partial_x u_\varepsilon)^3\partial_x^2 v_\varepsilon dx,\\
&-\varepsilon\int_{\mathbb{R}}\left(2a^2\partial_x^4 u_\varepsilon + Dagu_\varepsilon(\partial_x u_\varepsilon)^2 + Eagu_\varepsilon^2\partial_x^2 u_\varepsilon\right)\partial_x^4 u_\varepsilon dx\\
&\quad = -2a^2\varepsilon\left\|\partial_x^4 u_\varepsilon(t,\cdot)\right\|^2_{L^2(\mathbb{R})} + Dag\varepsilon\int_{\mathbb{R}} u_\varepsilon(\partial_x u_\varepsilon)^2\partial_x^4 u_\varepsilon dx\\
&\qquad + Eag\varepsilon\int_{\mathbb{R}} u_\varepsilon^2\partial_x^2 u_\varepsilon\partial_x^4 u_\varepsilon dx.
\end{aligned}$$

Consequently, an integration on $\mathbb{R}$ of (58) gives

$$\begin{aligned}
&a^2\frac{d}{dt}\left\|\partial_x^2 u_\varepsilon(t,\cdot)\right\|^2_{L^2(\mathbb{R})}+Dag\int_{\mathbb{R}}u_\varepsilon(\partial_x u_\varepsilon)^2\partial_t u_\varepsilon dx+Eag\int_{\mathbb{R}}u_\varepsilon^2\partial_x^2 u_\varepsilon\partial_t u_\varepsilon dx\\
&\quad+2a^2\varepsilon\left\|\partial_x^4 u_\varepsilon(t,\cdot)\right\|^2_{L^2(\mathbb{R})}\\
&\quad=-a^2g\,(30+2D+E)\int_{\mathbb{R}}u_\varepsilon\partial_x u_\varepsilon(\partial_x^2 u_\varepsilon)^2dx-(3D-6E)\,ag^2\int_{\mathbb{R}}u_\varepsilon^3(\partial_x u_\varepsilon)^3dx\\
&\quad-5a^2q\int_{\mathbb{R}}\partial_x v_\varepsilon(\partial_x^2 u_\varepsilon)^2dx-6a^2q\int_{\mathbb{R}}\partial_x^2 v_\varepsilon\partial_x u_\varepsilon\partial_x^2 u_\varepsilon dx\\
&\quad-2a^2q\int_{\mathbb{R}}u_\varepsilon\partial_x^3 v_\varepsilon\partial_x^2 u_\varepsilon dx-aq\,(D-3E)\int_{\mathbb{R}}u_\varepsilon^2(\partial_x u_\varepsilon)^2\partial_x v_\varepsilon dx\\
&\quad-(D-E)\,agq\int_{\mathbb{R}}u_\varepsilon v_\varepsilon(\partial_x u_\varepsilon)^3dx+\frac{Eagq}{2}\int_{\mathbb{R}}u_\varepsilon^2(\partial_x u_\varepsilon)^3\partial_x^2 v_\varepsilon dx\\
&\quad+2a^2b\int_{\mathbb{R}}P_\varepsilon\partial_x^4 u_\varepsilon dx+Dagb\int_{\mathbb{R}}P_\varepsilon u_\varepsilon(\partial_x u_\varepsilon)^2dx+Eagb\int_{\mathbb{R}}P_\varepsilon u_\varepsilon^2\partial_x^2 u_\varepsilon dx\\
&\quad-Dag\varepsilon\int_{\mathbb{R}}u_\varepsilon(\partial_x u_\varepsilon)^2\partial_x^4 u_\varepsilon dx+Eag\varepsilon\int_{\mathbb{R}}u_\varepsilon^2\partial_x^2 u_\varepsilon\partial_x^4 u_\varepsilon dx.
\end{aligned}\tag{59}$$

Thanks to the second equation of (16) and (18), we have that

$$\begin{aligned}
2a^2b\int_{\mathbb{R}}P_\varepsilon\partial_x^4 u_\varepsilon dx&=-2a^2b\int_{\mathbb{R}}\partial_x P_\varepsilon\partial_x^3 u_\varepsilon=-2a^2b\int_{\mathbb{R}}u_\varepsilon\partial_x^3 u_\varepsilon dx\\
&=2a^2b\int_{\mathbb{R}}\partial_x u_\varepsilon\partial_x^2 u_\varepsilon dx=0,\\
Eagb\int_{\mathbb{R}}P_\varepsilon u_\varepsilon^2\partial_x^2 u_\varepsilon dx&=-Eagb\int_{\mathbb{R}}\partial_x P_\varepsilon u_\varepsilon^2\partial_x u_\varepsilon dx-2Eagb\int_{\mathbb{R}}P_\varepsilon u_\varepsilon(\partial_x u_\varepsilon)^2dx\\
&=-2Eagb\int_{\mathbb{R}}P_\varepsilon u_\varepsilon(\partial_x u_\varepsilon)^2dx-Eagb\int_{\mathbb{R}}u_\varepsilon^3\partial_x u_\varepsilon dx\\
&=-2Eagb\int_{\mathbb{R}}P_\varepsilon u_\varepsilon(\partial_x u_\varepsilon)^2dx.
\end{aligned}$$

Therefore, by (59),

$$\begin{aligned}
&a^2\frac{d}{dt}\left\|\partial_x^2 u_\varepsilon(t,\cdot)\right\|^2_{L^2(\mathbb{R})}+Dag\int_{\mathbb{R}}u_\varepsilon(\partial_x u_\varepsilon)^2\partial_t u_\varepsilon dx+Eag\int_{\mathbb{R}}u_\varepsilon^2\partial_x^2 u_\varepsilon\partial_t u_\varepsilon dx\\
&\quad+2a^2\varepsilon\left\|\partial_x^4 u_\varepsilon(t,\cdot)\right\|^2_{L^2(\mathbb{R})}\\
&\quad=-a^2g\,(30+2D+E)\int_{\mathbb{R}}u_\varepsilon\partial_x u_\varepsilon(\partial_x^2 u_\varepsilon)^2dx-(3D-6E)\,ag^2\int_{\mathbb{R}}u_\varepsilon^3(\partial_x u_\varepsilon)^3dx\\
&\quad-5a^2q\int_{\mathbb{R}}\partial_x v_\varepsilon(\partial_x^2 u_\varepsilon)^2dx-6a^2q\int_{\mathbb{R}}\partial_x^2 v_\varepsilon\partial_x u_\varepsilon\partial_x^2 u_\varepsilon dx\\
&\quad-2a^2q\int_{\mathbb{R}}u_\varepsilon\partial_x^3 v_\varepsilon\partial_x^2 u_\varepsilon dx-aq\,(D-3E)\int_{\mathbb{R}}u_\varepsilon^2(\partial_x u_\varepsilon)^2\partial_x v_\varepsilon dx\\
&\quad-(D-E)\,agq\int_{\mathbb{R}}u_\varepsilon v_\varepsilon(\partial_x u_\varepsilon)^3dx+\frac{Eagq}{2}\int_{\mathbb{R}}u_\varepsilon^2(\partial_x u_\varepsilon)^3\partial_x^2 v_\varepsilon dx\\
&\quad+(D-2E)\,agb\int_{\mathbb{R}}P_\varepsilon u_\varepsilon(\partial_x u_\varepsilon)^2dx+Dag\varepsilon\int_{\mathbb{R}}u_\varepsilon(\partial_x u_\varepsilon)^2\partial_x^4 u_\varepsilon dx\\
&\quad+Eag\varepsilon\int_{\mathbb{R}}u_\varepsilon^2\partial_x^2 u_\varepsilon\partial_x^4 u_\varepsilon dx.
\end{aligned}\tag{60}$$

Observe that

$$
\begin{aligned}
&Dag\int_{\mathbb{R}} u_\varepsilon(\partial_x u_\varepsilon)^2\partial_t u_\varepsilon dx + Eag\int_{\mathbb{R}} u_\varepsilon^2\partial_x^2 u_\varepsilon\partial_t u_\varepsilon dx\\
&\quad=\frac{Dag}{2}\int_{\mathbb{R}}\partial_t(u_\varepsilon^2)(\partial_x u_\varepsilon)^2dx - Eag\int_{\mathbb{R}}\partial_x u_\varepsilon\partial_x(u_\varepsilon^2\partial_t u_\varepsilon)dx\\
&\quad=\frac{Dag}{2}\int_{\mathbb{R}}\partial_t(u_\varepsilon^2)(\partial_x u_\varepsilon)^2dx - 2Eag\int_{\mathbb{R}} u_\varepsilon(\partial_x u_\varepsilon)^2\partial_t u_\varepsilon dx - Ea\int_{\mathbb{R}} u_\varepsilon^2\partial_x u_\varepsilon\partial_{tx}^2 u_\varepsilon dx\\
&\quad ag\left(\frac{D}{2}-E\right)\int_{\mathbb{R}}\partial_t(u_\varepsilon^2)(\partial_x u_\varepsilon)^2dx - \frac{Eag}{2}\int_{\mathbb{R}} u_\varepsilon^2\partial_t((\partial_x u_\varepsilon)^2)dx.
\end{aligned}
$$

Consequently, by (60),

$$
\begin{aligned}
&a^2\frac{d}{dt}\left\|\partial_x^2 u_\varepsilon(t,\cdot)\right\|_{L^2(\mathbb{R})}^2 + ag\left(\frac{D}{2}-E\right)\int_{\mathbb{R}}\partial_t(u_\varepsilon^2)(\partial_x u_\varepsilon)^2dx\\
&\quad-\frac{Eag}{2}\int_{\mathbb{R}} u_\varepsilon^2\partial_t((\partial_x u_\varepsilon)^2)dx + 2a^2\varepsilon\left\|\partial_x^4 u_\varepsilon(t,\cdot)\right\|_{L^2(\mathbb{R})}^2\\
&\quad=-a^2g\,(30+2D+E)\int_{\mathbb{R}} u_\varepsilon\partial_x u_\varepsilon(\partial_x^2 u_\varepsilon)^2dx - (3D-6E)\,ag^2\int_{\mathbb{R}} u_\varepsilon^3(\partial_x u_\varepsilon)^3dx\\
&\quad-5a^2q\int_{\mathbb{R}}\partial_x v_\varepsilon(\partial_x^2 u_\varepsilon)^2dx - 6a^2q\int_{\mathbb{R}}\partial_x^2 v_\varepsilon\partial_x u_\varepsilon\partial_x^2 u_\varepsilon dx\\
&\quad-2a^2q\int_{\mathbb{R}} u_\varepsilon\partial_x^3 v_\varepsilon\partial_x^2 u_\varepsilon dx - aq\,(D-3E)\int_{\mathbb{R}} u_\varepsilon^2(\partial_x u_\varepsilon)^2\partial_x v_\varepsilon dx\\
&\quad-(D-E)\,agq\int_{\mathbb{R}} u_\varepsilon v_\varepsilon(\partial_x u_\varepsilon)^3dx + \frac{Eagq}{2}\int_{\mathbb{R}} u_\varepsilon^2(\partial_x u_\varepsilon)^3\partial_x^2 v_\varepsilon dx\\
&\quad+(D-2E)\,agb\int_{\mathbb{R}} P_\varepsilon u_\varepsilon(\partial_x u_\varepsilon)^2dx + Dag\varepsilon\int_{\mathbb{R}} u_\varepsilon(\partial_x u_\varepsilon)^2\partial_x^4 u_\varepsilon dx\\
&\quad+Eag\varepsilon\int_{\mathbb{R}} u_\varepsilon^2\partial_x^2 u_\varepsilon\partial_x^4 u_\varepsilon dx.
\end{aligned}
\tag{61}
$$

We search $D$, $E$ such that

$$
\frac{D}{2}-E=-\frac{E}{2},\qquad 30+2D+E=0,
$$

that is

$$
D=E,\qquad 30+2D+E=0. \tag{62}
$$

Since $D=E-10$ is the unique solution of (62), it follows from (61) that

$$
\begin{aligned}
&a^2\frac{d}{dt}\left\|\partial_x^2 u_\varepsilon(t,\cdot)\right\|_{L^2(\mathbb{R})}^2 + 5ag\int_{\mathbb{R}}\partial_t(u_\varepsilon^2)(\partial_x u_\varepsilon)^2dx + 5ag\int_{\mathbb{R}} u_\varepsilon^2\partial_t((\partial_x u_\varepsilon)^2)dx\\
&\quad+2a^2\varepsilon\left\|\partial_x^4 u_\varepsilon(t,\cdot)\right\|_{L^2(\mathbb{R})}^2\\
&\quad=-30ag^2\int_{\mathbb{R}} u_\varepsilon^3(\partial_x u_\varepsilon)^3dx - 5a^2q\int_{\mathbb{R}}\partial_x v_\varepsilon(\partial_x^2 u_\varepsilon)^2dx - 6a^2q\int_{\mathbb{R}}\partial_x^2 v_\varepsilon\partial_x u_\varepsilon\partial_x^2 u_\varepsilon dx\\
&\quad-2a^2q\int_{\mathbb{R}} u_\varepsilon\partial_x^3 v_\varepsilon\partial_x^2 u_\varepsilon dx - 20aq\int_{\mathbb{R}} u_\varepsilon^2(\partial_x u_\varepsilon)^2\partial_x v_\varepsilon dx - 5agq\int_{\mathbb{R}} u_\varepsilon^2(\partial_x u_\varepsilon)^3\partial_x^2 v_\varepsilon dx\\
&\quad+10agb\int_{\mathbb{R}} P_\varepsilon u_\varepsilon(\partial_x u_\varepsilon)^2dx - 10ag\varepsilon\int_{\mathbb{R}} u_\varepsilon(\partial_x u_\varepsilon)^2\partial_x^4 u_\varepsilon dx\\
&\quad-10ag\varepsilon\int_{\mathbb{R}} u_\varepsilon^2\partial_x^2 u_\varepsilon\partial_x^4 u_\varepsilon dx,
\end{aligned}
$$

that is

$$\begin{aligned}
&\frac{d}{dt}\left(a^2\left\|\partial_x^2 u_\varepsilon(t,\cdot)\right\|^2_{L^2(\mathbb{R})}+5ag\int_{\mathbb{R}}u_\varepsilon^2(\partial_x u_\varepsilon)^2dx\right)+2a^2\varepsilon\left\|\partial_x^4 u_\varepsilon(t,\cdot)\right\|^2_{L^2(\mathbb{R})}\\
&\quad=-30ag^2\int_{\mathbb{R}}u_\varepsilon^3(\partial_x u_\varepsilon)^3dx-5a^2q\int_{\mathbb{R}}\partial_x v_\varepsilon(\partial_x^2 u_\varepsilon)^2dx\\
&\quad-6a^2q\int_{\mathbb{R}}\partial_x^2 v_\varepsilon\partial_x u_\varepsilon\partial_x^2 u_\varepsilon dx-2a^2q\int_{\mathbb{R}}u_\varepsilon\partial_x^3 v_\varepsilon\partial_x^2 u_\varepsilon dx\\
&\quad-20aq\int_{\mathbb{R}}u_\varepsilon^2(\partial_x u_\varepsilon)^2\partial_x v_\varepsilon dx-5agq\int_{\mathbb{R}}u_\varepsilon^2(\partial_x u_\varepsilon)^3\partial_x^2 v_\varepsilon dx\\
&\quad+10agb\int_{\mathbb{R}}P_\varepsilon u_\varepsilon(\partial_x u_\varepsilon)^2dx-10ag\varepsilon\int_{\mathbb{R}}u_\varepsilon(\partial_x u_\varepsilon)^2\partial_x^4 u_\varepsilon dx\\
&\quad-10ag\varepsilon\int_{\mathbb{R}}u_\varepsilon^2\partial_x^2 u_\varepsilon\partial_x^4 u_\varepsilon dx.
\end{aligned} \tag{63}$$

Due to (41), (42), (43), (55), Lemma 3 and the Young inequality,

$$\begin{aligned}
|30ag^2|\int_{\mathbb{R}}|u_\varepsilon|^3|\partial_x u_\varepsilon|^3dx\le&|30ag^2|\left\|u_\varepsilon\right\|^3_{L^\infty((0,T)\times\mathbb{R})}\int_{\mathbb{R}}|\partial_x u_\varepsilon|^3dx\\
\le&C(T)\int_{\mathbb{R}}|\partial_x u_\varepsilon|^3dx\\
\le&C(T)\left\|\partial_x u_\varepsilon(t,\cdot)\right\|^2_{L^2(\mathbb{R})}+C(T)\int_{\mathbb{R}}(\partial_x u_\varepsilon)^4dx\\
\le&C(T)+C(T)\left\|\partial_x u_\varepsilon\right\|^2_{L^\infty((0,T)\times\mathbb{R})}\left\|\partial_x u_\varepsilon(t,\cdot)\right\|^2_{L^2(\mathbb{R})}\\
\le&C(T)\left(1+\left\|\partial_x u_\varepsilon\right\|^2_{L^\infty((0,T)\times\mathbb{R})}\right),\\
|5a^2q|\int_{\mathbb{R}}|\partial_x v_\varepsilon|(\partial_x^2 u_\varepsilon)^2dx\le&|5a^2q|\left\|\partial_x v_\varepsilon(t,\cdot)\right\|_{L^\infty(\mathbb{R})}\left\|\partial_x^2 u_\varepsilon(t,\cdot)\right\|^2_{L^2(\mathbb{R})}\\
\le&C_0\left\|\partial_x^2 u_\varepsilon(t,\cdot)\right\|^2_{L^2(\mathbb{R})},\\
|6a^2q|\int_{\mathbb{R}}|\partial_x^2 v_\varepsilon\partial_x u_\varepsilon||\partial_x^2 u_\varepsilon|dx\le&3a^4q^2\int_{\mathbb{R}}(\partial_x^2 v_\varepsilon)^2(\partial_x u_\varepsilon)^2dx+3\left\|\partial_x^2 u_\varepsilon(t,\cdot)\right\|^2_{L^2(\mathbb{R})}\\
\le&3a^4q^2\left\|\partial_x^2 v_\varepsilon\right\|^2_{L^\infty((0,T)\times\mathbb{R})}\left\|\partial_x u_\varepsilon(t,\cdot)\right\|^2_{L^2(\mathbb{R})}+3\left\|\partial_x^2 u_\varepsilon(t,\cdot)\right\|^2_{L^2(\mathbb{R})}\\
\le&C(T)+3\left\|\partial_x^2 u_\varepsilon(t,\cdot)\right\|^2_{L^2(\mathbb{R})},\\
|2a^2q|\int_{\mathbb{R}}|u_\varepsilon\partial_x^3 v_\varepsilon||\partial_x^2 u_\varepsilon|dx\le&a^4q^2\int_{\mathbb{R}}u_\varepsilon^2(\partial_x^3 v_\varepsilon)^2dx+\left\|\partial_x^2 u_\varepsilon(t,\cdot)\right\|^2_{L^2(\mathbb{R})}\\
\le&a^4q^2\left\|u_\varepsilon\right\|^2_{L^\infty((0,T)\times\mathbb{R})}\left\|\partial_x^3 v_\varepsilon(t,\cdot)\right\|^2_{L^2(\mathbb{R})}+\left\|\partial_x^2 u_\varepsilon(t,\cdot)\right\|^2_{L^2(\mathbb{R})}\\
\le&C(T)+\left\|\partial_x^2 u_\varepsilon(t,\cdot)\right\|^2_{L^2(\mathbb{R})},\\
|20aq|\int_{\mathbb{R}}u_\varepsilon^2(\partial_x u_\varepsilon)^2|\partial_x v_\varepsilon|dx\le&|20aq|\left\|\partial_x v_\varepsilon(t,\cdot)\right\|_{L^\infty(\mathbb{R})}\left\|u_\varepsilon\right\|^2_{L^\infty((0,T)\times\mathbb{R})}\left\|\partial_x u_\varepsilon(t,\cdot)\right\|^2_{L^2(\mathbb{R})}\\
\le&C(T)\left\|\partial_x u_\varepsilon(t,\cdot)\right\|^2_{L^2(\mathbb{R})}\le C(T),
\end{aligned}$$

$$
\begin{aligned}
|5agq|\int_{\mathbb{R}} u_\varepsilon^2|\partial_x u_\varepsilon|^3|\partial_x^2 v_\varepsilon|dx \leq&|5agq|\left\|u_\varepsilon\right\|^2_{L^\infty((0,T)\times\mathbb{R})}\left\|\partial_x^2 v_\varepsilon\right\|_{L^\infty((0,T)\times\mathbb{R})}\int_{\mathbb{R}}|\partial_x u_\varepsilon|^3dx\\
\leq&C(T)\int_{\mathbb{R}}|\partial_x u_\varepsilon|^3dx\\
\leq&C(T)\left\|\partial_x u_\varepsilon(t,\cdot)\right\|^2_{L^2(\mathbb{R})}+C(T)\int_{\mathbb{R}}(\partial_x u_\varepsilon)^4dx\\
\leq&C(T)+C(T)\left\|\partial_x u_\varepsilon\right\|^2_{L^\infty((0,T)\times\mathbb{R})}\left\|\partial_x u_\varepsilon(t,\cdot)\right\|^2_{L^2(\mathbb{R})}\\
\leq&C(T)\left(1+\left\|\partial_x u_\varepsilon\right\|^2_{L^\infty((0,T)\times\mathbb{R})}\right),\\
|10agb|\int_{\mathbb{R}}|P_\varepsilon u_\varepsilon|(\partial_x u_\varepsilon)^2dx \leq&|10agb|\left\|P_\varepsilon\right\|_{L^\infty((0,T)\times\mathbb{R})}\left\|u_\varepsilon\right\|_{L^\infty((0,T)\times\mathbb{R})}\left\|\partial_x u_\varepsilon(t,\cdot)\right\|^2_{L^2(\mathbb{R})}\\
\leq&C(T)\left\|\partial_x u_\varepsilon(t,\cdot)\right\|^2_{L^2(\mathbb{R})}\leq C(T),\\
|10ag|\varepsilon\int_{\mathbb{R}}|u_\varepsilon(\partial_x u_\varepsilon)^2||\partial_x^4 u_\varepsilon|dx =&\varepsilon\int_{\mathbb{R}}|10gu_\varepsilon(\partial_x u_\varepsilon)^2||a\partial_x^4 u_\varepsilon|dx\\
\leq&50g^2\varepsilon\int_{\mathbb{R}}u_\varepsilon^2(\partial_x u_\varepsilon)^4dx+\frac{a^2\varepsilon}{2}\left\|\partial_x^4 u_\varepsilon(t,\cdot)\right\|^2_{L^2(\mathbb{R})}\\
\leq&50g^2\varepsilon\left\|u_\varepsilon\right\|^2_{L^\infty((0,T)\times\mathbb{R})}\int_{\mathbb{R}}(\partial_x u_\varepsilon)^4dx+\frac{a^2\varepsilon}{2}\left\|\partial_x^4 u_\varepsilon(t,\cdot)\right\|^2_{L^2(\mathbb{R})}\\
\leq&C(T)\varepsilon\left\|\partial_x u_\varepsilon\right\|^2_{L^\infty((0,T)\times\mathbb{R})}\left\|\partial_x u_\varepsilon(t,\cdot)\right\|^2_{L^2(\mathbb{R})}\\
&+\frac{a^2\varepsilon}{2}\left\|\partial_x^4 u_\varepsilon(t,\cdot)\right\|^2_{L^2(\mathbb{R})},\\
|10ag|\varepsilon\int_{\mathbb{R}}|u_\varepsilon^2\partial_x^2 u_\varepsilon||\partial_x^4 u_\varepsilon|dx =&\varepsilon\int_{\mathbb{R}}|10gu_\varepsilon^2\partial_x^2 u_\varepsilon||a\partial_x^4 u_\varepsilon|dx\\
\leq&50g^2\varepsilon\int_{\mathbb{R}}u_\varepsilon^4(\partial_x^2 u_\varepsilon)^2dx+\frac{a^2\varepsilon}{2}\left\|\partial_x^4 u_\varepsilon(t,\cdot)\right\|^2_{L^2(\mathbb{R})}\\
\leq&50g^2\varepsilon\left\|u_\varepsilon\right\|^4_{L^\infty((0,T)\times\mathbb{R})}\left\|\partial_x^2 u_\varepsilon(t,\cdot)\right\|^2_{L^2(\mathbb{R})}+\frac{a^2\varepsilon}{2}\left\|\partial_x^4 u_\varepsilon(t,\cdot)\right\|^2_{L^2(\mathbb{R})}\\
\leq&C(T)\varepsilon\left\|\partial_x^2 u_\varepsilon(t,\cdot)\right\|^2_{L^2(\mathbb{R})}+\frac{a^2\varepsilon}{2}\left\|\partial_x^4 u_\varepsilon(t,\cdot)\right\|^2_{L^2(\mathbb{R})}.
\end{aligned}
$$

Therefore, defining

$$
G_1(t)=a^2\left\|\partial_x^2 u_\varepsilon(t,\cdot)\right\|^2_{L^2(\mathbb{R})}+5ag\int_{\mathbb{R}}u_\varepsilon^2(\partial_x u_\varepsilon)^2dx, \tag{64}
$$

by (63) and (64), we have

$$
\begin{aligned}
\frac{dG_1(t)}{dt}+\varepsilon\left\|\partial_x^4 u_\varepsilon(t,\cdot)\right\|^2_{L^2(\mathbb{R})}\leq&C_0\left\|\partial_x^2 u_\varepsilon(t,\cdot)\right\|^2_{L^2(\mathbb{R})}+C(T)\left(1+\left\|\partial_x u_\varepsilon\right\|^2_{L^\infty((0,T)\times\mathbb{R})}\right)\\
&+C(T)\varepsilon\left\|\partial_x u_\varepsilon\right\|^2_{L^\infty((0,T)\times\mathbb{R})}\left\|\partial_x u_\varepsilon(t,\cdot)\right\|^2_{L^2(\mathbb{R})}\\
&+C(T)\varepsilon\left\|\partial_x^2 u_\varepsilon(t,\cdot)\right\|^2_{L^2(\mathbb{R})}.
\end{aligned} \tag{65}
$$

Observe that by (41), (42) and (64),

$$\begin{aligned}
C_0\left\|\partial_x^2 u_\varepsilon(t,\cdot)\right\|^2_{L^2(\mathbb{R})} =&\frac{C_0a^2}{a^2}\left\|\partial_x^2 u_\varepsilon(t,\cdot)\right\|^2_{L^2(\mathbb{R})}\\
=&\frac{C_0}{a^2}G_1(t)-\frac{5C_0g}{a}\int_{\mathbb{R}}u_\varepsilon^2(\partial_x u_\varepsilon)^2dx\\
\le&C_0G_1(t)+\left|\frac{5C_0g}{a}\right|\int_{\mathbb{R}}u_\varepsilon^2(\partial_x u_\varepsilon)^2dx\\
\le&C_0G_1(t)+\left|\frac{5C_0g}{a}\right|\left\|u_\varepsilon\right\|^2_{L^\infty((0,T)\times\mathbb{R})}\left\|\partial_x u_\varepsilon(t,\cdot)\right\|^2_{L^2(\mathbb{R})}\\
\le&C_0G_1(t)+C(T).
\end{aligned}\tag{66}$$

It follows from (65) and (66) that

$$\begin{aligned}
\frac{dG_1(t)}{dt}+\varepsilon\left\|\partial_x^4 u_\varepsilon(t,\cdot)\right\|^2_{L^2(\mathbb{R})}\le&C_0G_1(t)+C(T)\left(1+\left\|\partial_x u_\varepsilon\right\|^2_{L^\infty((0,T)\times\mathbb{R})}\right)\\
&+C(T)\varepsilon\left\|\partial_x u_\varepsilon\right\|^2_{L^\infty((0,T)\times\mathbb{R})}\left\|\partial_x u_\varepsilon(t,\cdot)\right\|^2_{L^2(\mathbb{R})}\\
&+C(T)\varepsilon\left\|\partial_x^2 u_\varepsilon(t,\cdot)\right\|^2_{L^2(\mathbb{R})}.
\end{aligned}$$

The Gronwall Lemma, (17) and Lemma 3 give

$$\begin{aligned}
&\left\|\partial_x^2 u_\varepsilon(t,\cdot)\right\|^2_{L^2(\mathbb{R})}+5ag\int_{\mathbb{R}}u_\varepsilon^2(\partial_x u_\varepsilon)^2dx+2\varepsilon e^{C_0t}\int_0^t e^{-C_0s}\left\|\partial_x^4 u_\varepsilon(s,\cdot)\right\|^2_{L^2(\mathbb{R})}ds\\
&\quad\le C_0e^{C_0t}+C(T)\left(1+\left\|\partial_x u_\varepsilon\right\|^2_{L^\infty((0,T)\times\mathbb{R})}\right)e^{C_0t}\int_0^t e^{-C_0s}ds\\
&\qquad+C(T)\varepsilon\left\|\partial_x u_\varepsilon\right\|^2_{L^\infty((0,T)\times\mathbb{R})}e^{C_0t}\int_0^t e^{-C_0s}\left\|\partial_x u_\varepsilon(s,\cdot)\right\|^2_{L^2(\mathbb{R})}ds\\
&\qquad+C(T))\varepsilon e^{C_0t}\int_0^t e^{-C_0s}\left\|\partial_x^2 u_\varepsilon(s,\cdot)\right\|^2_{L^2(\mathbb{R})}ds\\
&\quad\le C(T)\left(1+\left\|\partial_x u_\varepsilon\right\|^2_{L^\infty((0,T)\times\mathbb{R})}\right)\\
&\qquad+C(T)\varepsilon\left\|\partial_x u_\varepsilon\right\|^2_{L^\infty((0,T)\times\mathbb{R})}\int_0^t\left\|\partial_x u_\varepsilon(s,\cdot)\right\|^2_{L^2(\mathbb{R})}ds\\
&\qquad+C(T))\varepsilon\int_0^t\left\|\partial_x^2 u_\varepsilon(s,\cdot)\right\|^2_{L^2(\mathbb{R})}ds\\
&\quad\le C(T)\left(1+\left\|\partial_x u_\varepsilon\right\|^2_{L^\infty((0,T)\times\mathbb{R})}\right).
\end{aligned}$$

Therefore, thanks to (41) and (42),

$$\begin{aligned}
&\left\|\partial_x^2 u_\varepsilon(t,\cdot)\right\|^2_{L^2(\mathbb{R})}+2\varepsilon e^{C_0t}\int_0^t e^{-C_0s}\left\|\partial_x^4 u_\varepsilon(s,\cdot)\right\|^2_{L^2(\mathbb{R})}ds\\
&\quad=C(T)\left(1+\left\|\partial_x u_\varepsilon\right\|^2_{L^\infty((0,T)\times\mathbb{R})}\right)-5ag\int_{\mathbb{R}}u_\varepsilon^2(\partial_x u_\varepsilon)^2dx\\
&\quad\le C(T)\left(1+\left\|\partial_x u_\varepsilon\right\|^2_{L^\infty((0,T)\times\mathbb{R})}\right)+|5ag|\int_{\mathbb{R}}u_\varepsilon^2(\partial_x u_\varepsilon)^2dx\\
&\quad\le C(T)\left(1+\left\|\partial_x u_\varepsilon\right\|^2_{L^\infty((0,T)\times\mathbb{R})}\right)+|5ag|\left\|u_\varepsilon\right\|^2_{L^\infty((0,T)\times\mathbb{R})}\left\|\partial_x u_\varepsilon(t,\cdot)\right\|^2_{L^2(\mathbb{R})}\\
&\quad\le C(T)\left(1+\left\|\partial_x u_\varepsilon\right\|^2_{L^\infty((0,T)\times\mathbb{R})}\right).
\end{aligned}\tag{67}$$

We prove (56). Due to (42), (67) and the Hölder inequality,

$$\begin{aligned}(\partial_x u_\varepsilon(t,x))^2 =&2\int_{-\infty}^{x} \partial_x u_\varepsilon \partial_x^2 u_\varepsilon dx \leq 2\int_{\mathbb{R}} |\partial_x u_\varepsilon||\partial_x^2 u_\varepsilon| dx\\ &\leq \|\partial_x u_\varepsilon(t,\cdot)\|_{L^2(\mathbb{R})} \left\|\partial_x^2 u_\varepsilon(t,\cdot)\right\|_{L^2(\mathbb{R})} \leq C(T)\sqrt{\left(1+\|\partial_x u_\varepsilon\|^2_{L^\infty((0,T)\times\mathbb{R})}\right)}.\end{aligned}$$

Therefore,

$$\|\partial_x u_\varepsilon\|^4_{L^\infty((0,T)\times\mathbb{R})} - C(T)\,\|\partial_x u_\varepsilon\|^2_{L^\infty((0,T)\times\mathbb{R})} - C(T) \leq 0,$$

which gives (56).

Finally, (57) follows from (56) and (67). □

**Lemma 10.** *Assume* (5)*. Fix $T > 0$. There exists a constant $C(T) > 0$, independent on $\varepsilon$, such that*

$$\left\|\partial_x^4 v_\varepsilon(t,\cdot)\right\|^2_{L^2(\mathbb{R})} \leq C(T), \tag{68}$$

*for every $0 \leq t \leq T$. In particular, we have that*

$$\left\|\partial_x^3 v_\varepsilon\right\|_{L^\infty((0,T)\times\mathbb{R})} \leq C(T), \tag{69}$$

**Proof.** Let $0 \leq t \leq T$. Differentiating the third equation of (16) twice with respect to $x$, we have

$$\alpha\partial_x^4 v_\varepsilon = 2\kappa(\partial_x u_\varepsilon)^2 + 2\kappa u_\varepsilon \partial_x^2 u_\varepsilon - \beta\partial_x^3 v_\varepsilon - \gamma\partial_x^2 v_\varepsilon. \tag{70}$$

Since

$$u_\varepsilon(t,\pm\infty) = \partial_x u_\varepsilon(t,\pm\infty) = \partial_x^2 u_\varepsilon(t,\pm\infty) = 0, \tag{71}$$

it follows from (24) and (55) that

$$\partial_x^4 v_\varepsilon(t,\pm\infty) = 0. \tag{72}$$

Multiplying (70) by $2\alpha\partial_x^4 v_\varepsilon$, an integration on $\mathbb{R}$ gives

$$\begin{aligned}2\alpha^2\left\|\partial_x^4 v_\varepsilon(t,\cdot)\right\|^2_{L^2(\mathbb{R})} =&2\kappa\alpha\int_{\mathbb{R}}(\partial_x u_\varepsilon)^2\partial_x^4 v_\varepsilon dx + 2\kappa\alpha\int_{\mathbb{R}} u_\varepsilon\partial_x^2 u_\varepsilon \partial_x^4 v_\varepsilon dx\\ &- 2\beta\alpha\int_{\mathbb{R}}\partial_x^3 v_\varepsilon\partial_x^4 v_\varepsilon dx - 2\gamma\alpha\int_{\mathbb{R}}\partial_x^2 v_\varepsilon \partial_x^4 v_\varepsilon dx.\end{aligned} \tag{73}$$

Observe that, thanks to (24), (55) and (72),

$$\begin{aligned}-2\beta\alpha\int_{\mathbb{R}}\partial_x^3 v_\varepsilon \partial_x^4 v_\varepsilon dx =&0,\\ -2\gamma\alpha\int_{\mathbb{R}}\partial_x^2 v_\varepsilon\partial_x^4 v_\varepsilon dx =&2\gamma\alpha\left\|\partial_x^3 v_\varepsilon(t,\cdot)\right\|^2_{L^2(\mathbb{R})}.\end{aligned}$$

Therefore, by (55) and (73),

$$\begin{aligned}2\alpha^2\left\|\partial_x^4 v_\varepsilon(t,\cdot)\right\|^2_{L^2(\mathbb{R})} \leq&2|\kappa\alpha|\int_{\mathbb{R}}(\partial_x u_\varepsilon)^2|\partial_x^4 v_\varepsilon| dx + 2|\kappa\alpha|\int_{\mathbb{R}}|u_\varepsilon\partial_x^2 u_\varepsilon||\partial_x^4 v_\varepsilon| dx\\ &+ 2|\gamma\alpha|\left\|\partial_x^3 v_\varepsilon(t,\cdot)\right\|^2_{L^2(\mathbb{R})}\\ \leq&2|\kappa\alpha|\int_{\mathbb{R}}(\partial_x u_\varepsilon)^2|\partial_x^4 v_\varepsilon| dx + 2|\kappa\alpha|\int_{\mathbb{R}}|u_\varepsilon\partial_x^2 u_\varepsilon||\partial_x^4 v_\varepsilon| dx + C(T).\end{aligned} \tag{74}$$

Due to (41), (42), (56), (57) and the Young inequality,

$$\begin{aligned}
2|\kappa\alpha|\int_{\mathbb{R}}(\partial_x u_\varepsilon)^2|\partial_x^4 v_\varepsilon|dx \leq&\kappa^2\int_{\mathbb{R}}(\partial_x u_\varepsilon)^4dx+\alpha^2\left\|\partial_x^4 v_\varepsilon(t,\cdot)\right\|^2_{L^2(\mathbb{R})}\\
\leq&\kappa^2\left\|\partial_x u_\varepsilon\right\|^2_{L^\infty((0,T)\times\mathbb{R})}\left\|\partial_x u_\varepsilon(t,\cdot)\right\|^2_{L^2(\mathbb{R})}+\alpha^2\left\|\partial_x^4 v_\varepsilon(t,\cdot)\right\|^2_{L^2(\mathbb{R})}\\
\leq&C(T)+\alpha^2\left\|\partial_x^4 v_\varepsilon(t,\cdot)\right\|^2_{L^2(\mathbb{R})},\\
2|\kappa\alpha|\int_{\mathbb{R}}|u_\varepsilon\partial_x^2 u_\varepsilon||\partial_x^4 v_\varepsilon|=&\int_{\mathbb{R}}|2\kappa u_\varepsilon\partial_x^2 u_\varepsilon||\alpha\partial_x^4 v_\varepsilon|dx\\
\leq&2\kappa^2\int_{\mathbb{R}}u_\varepsilon^2(\partial_x^2 u_\varepsilon)^2dx+\frac{\alpha^2}{2}\left\|\partial_x^4 v_\varepsilon(t,\cdot)\right\|^2_{L^2(\mathbb{R})}\\
\leq&2\kappa^2\left\|u_\varepsilon\right\|^2_{L^\infty((0,T)\times\mathbb{R})}\left\|\partial_x^2 u_\varepsilon(t,\cdot)\right\|^2_{L^2(\mathbb{R})}+\frac{\alpha^2}{2}\left\|\partial_x^4 v_\varepsilon(t,\cdot)\right\|^2_{L^2(\mathbb{R})}\\
\leq&C(T)+\frac{\alpha^2}{2}\left\|\partial_x^4 v_\varepsilon(t,\cdot)\right\|^2_{L^2(\mathbb{R})}.
\end{aligned}$$

Consequently, by (74),

$$\frac{\alpha^2}{2}\left\|\partial_x^4 v_\varepsilon(t,\cdot)\right\|^2_{L^2(\mathbb{R})}\leq C(T),$$

which gives (68).

Finally, we prove (69). Due to (55), (68) and the Hölder inequality,

$$\begin{aligned}
(\partial_x^3 v_\varepsilon(t,x))^2=&2\int_{-\infty}^{x}\partial_x^3 v_\varepsilon\partial_x^4 v_\varepsilon dx\leq 2\int_{\mathbb{R}}|\partial_x^3 v_\varepsilon||\partial_x^4 v_\varepsilon|dx\\
\leq&\left\|\partial_x^3 v_\varepsilon(t,\cdot)\right\|_{L^2(\mathbb{R})}\left\|\partial_x^4 v_\varepsilon(t,\cdot)\right\|_{L^2(\mathbb{R})}\leq C(T).
\end{aligned}$$

Hence,

$$\left\|\partial_x^3 v_\varepsilon\right\|^2_{L^\infty((0,T)\times\mathbb{R})}\leq C(T),$$

which gives (69). □

## 3. Proof of Theorem 1

This section is devoted to the proof of Theorem 1.

We begin by proving the following lemma.

**Lemma 11.** *Fix $T>0$. Then,*

$$\text{the sequence } \{u_\varepsilon\}_{\varepsilon>0} \text{ is compact in } L^2_{loc}((0,\infty)\times\mathbb{R}). \tag{75}$$

*Consequently, there exists a subsequence $\{u_{\varepsilon_k}\}_{k\in\mathbb{N}}$ of $\{u_\varepsilon\}_{\varepsilon>0}$ and $u\in L^2_{loc}((0,\infty)\times\mathbb{R})$ such that, for each compact subset $K$ of $(0,\infty)\times\mathbb{R})$,*

$$u_{\varepsilon_k}\to u \text{ in } L^2(K) \text{ and a.e.}, \tag{76}$$

$$v_{\varepsilon_k}\rightharpoonup v \text{ in } H^1((0,T)\times\mathbb{R}), \tag{77}$$

$$P_{\varepsilon_k}\rightharpoonup P \text{ in } L^2((0,T)\times\mathbb{R}). \tag{78}$$

*Moreover, $(u,v,P)$ is a solution of (1) satisfying (11) and (12).*

**Proof.** We begin by proving (75). To prove (75), we rely on the Aubin–Lions Lemma (see [58–60]). We recall that

$$H^1_{loc}(\mathbb{R}) \hookrightarrow\hookrightarrow L^2_{loc}(\mathbb{R}) \hookrightarrow H^{-1}_{loc}(\mathbb{R}),$$

where the first inclusion is compact and the second is continuous. Owing to the Aubin–Lions Lemma [60], to prove (75), it suffices to show that

$$\{u_\varepsilon\}_{\varepsilon>0} \text{ is uniformly bounded in } L^2(0,T;H^1_{loc}(\mathbb{R})), \tag{79}$$
$$\{\partial_t u_\varepsilon\}_{\varepsilon>0} \text{ is uniformly bounded in } L^2(0,T;H^{-1}_{loc}(\mathbb{R})). \tag{80}$$

We prove (79). Thanks to (42), (57) and Lemma 3,

$$\|u_\varepsilon(t,\cdot)\|^2_{H^2(\mathbb{R})} = \|u_\varepsilon(t,\cdot)\|^2_{L^2(\mathbb{R})} + \|\partial_x u_\varepsilon(t,\cdot)\|^2_{L^2(\mathbb{R})} + \left\|\partial_x^2 u_\varepsilon(t,\cdot)\right\|^2_{L^2(\mathbb{R})} \le C(T).$$

Therefore,

$$\{u_\varepsilon\}_{\varepsilon>0} \text{ is uniformly bounded in } L^\infty(0,T;H^2(\mathbb{R})),$$

which gives (79).

We prove (80). By the first equation of (16),

$$\partial_t u_\varepsilon = \partial_x\left(-gu_\varepsilon^3 + a\partial_x^2 u_\varepsilon - qv_\varepsilon u_\varepsilon - \varepsilon\partial_x^3 u_\varepsilon\right) + bP_\varepsilon. \tag{81}$$

We have that

$$\|u_\varepsilon\|_{L^6((0,T)\times\mathbb{R})} \le C(T). \tag{82}$$

Indeed, thanks to (41) and Lemma 3,

$$\begin{aligned} g^2\int_0^T\int_{\mathbb{R}} u_\varepsilon^6 dtdx \le& g^2\|u_\varepsilon\|^4_{L^\infty((0,T)\times\mathbb{R})}\int_0^T\int_{\mathbb{R}} u_\varepsilon^2 dtdx\\ \le& C(T)\int_0^T\int_{\mathbb{R}} u_\varepsilon^2 dtdx \le C(T). \end{aligned}$$

We prove that

$$q^2\|v_\varepsilon u_\varepsilon\|^2_{L^2((0,T)\times\mathbb{R})} \le C(T). \tag{83}$$

Due to Lemma 3,

$$\begin{aligned} q^2\int_0^T\int_{\mathbb{R}} v_\varepsilon^2 u_\varepsilon^2 dtdx \le& q^2\|v_\varepsilon\|^2_{L^\infty((0,T)\times\mathbb{R})}\int_0^T\int_{\mathbb{R}} u_\varepsilon^2 dtdx\\ \le& C(T)\int_0^T\int_{\mathbb{R}} u_\varepsilon^2 dtdx \le C(T). \end{aligned}$$

Observe that, since $0<\varepsilon<1$, thanks to (42) and (57),

$$\varepsilon\left\|\partial_x^3 u_\varepsilon\right\|^2_{L^2((0,T)\times\mathbb{R})},\ \beta^2\left\|\partial_x^2 u_\varepsilon\right\|^2_{L^2((0,T)\times\mathbb{R})} \le C(T). \tag{84}$$

Therefore, by (82), (83) and (84),

$$\left\{\partial_x\left(-gu_\varepsilon^3 + a\partial_x^2 u_\varepsilon - qv_\varepsilon u_\varepsilon - \varepsilon\partial_x^3 u_\varepsilon\right)\right\}_{\varepsilon>0} \text{ is bounded in } H^1((0,T)\times\mathbb{R}). \tag{85}$$

Moreover, by (43), we have that

$$b^2\|P_\varepsilon\|^2_{L^2((0,T)\times\mathbb{R})} \le C(T). \tag{86}$$

Equation (80) follows from (85) and (86).

Thanks to the Aubin–Lions Lemma, (75) and (76) hold.

Observe that, (77) follows from Lemma 3, while, by (43), we have (78). Consequently, $(u, v, P)$ solves (1).

Observe again that, thanks to Lemmas 3, 7, 8, 9, (10) and the second equation of (16), we obtain (11).

Finally, we prove (12). Thanks to Lemmas 3 and 7, we have

$$u_{\varepsilon_k} \rightharpoonup u \text{ in } H^1((0,T)\times\mathbb{R}). \tag{87}$$

Therefore, (12) follows from (19) and (87). □

We are ready for the proof of Theorem 1.

**Proof of Theorem 1.** Lemma 11 gives the existence of a solution of (1) satisfying (11) and (12). Let $(u_1, P_1)$ and $(u_2, P_2)$ be two solutions of (1) satisfying (11) and (12), namely

$$\begin{cases} \partial_t u_1 + 3gu_1^2\partial_x u_1 - a\partial_x^3 u_1 + q\partial_x(u_1v_1) = bP_1, & t>0,\ x\in\mathbb{R},\\ \partial_x P_1 = u_1, & t>0,\ x\in\mathbb{R},\\ \alpha\partial_x^2 v_1 + \beta\partial_x v_1 + \gamma v_1 = \kappa u_1^2, & t>0,\ x\in\mathbb{R},\\ P_1(t,-\infty) = 0, & t>0,\\ u_1(0,x) = u_{1,0}(x), & x\in\mathbb{R}, \end{cases}$$

$$\begin{cases} \partial_t u_2 + 3gu_2^2\partial_x u_2 - a\partial_x^3 u_2 + q\partial_x(u_2v_2) = bP_2, & t>0,\ x\in\mathbb{R},\\ \partial_x P_2 = u_2, & t>0,\ x\in\mathbb{R},\\ \alpha\partial_x^2 v_2 + \beta\partial_x v_2 + \gamma v_2 = \kappa u_2^2, & t>0,\ x\in\mathbb{R},\\ P_2(t,-\infty) = 0, & t>0,\\ u_2(0,x) = u_{2,0}(x), & x\in\mathbb{R}. \end{cases}$$

Then, the triad $(\omega, V, \Omega)$ defined by

$$\begin{aligned} &\omega(t,x) = u_1(t,x) - u_2(t,x), \quad V(t,x) = v_1(t,x) - v_2(t,x),\\ &\Omega(t,x) = \int_{-\infty}^{x}\omega(t,y)dy = \int_{-\infty}^{x}u_1(t,y)dy - \int_{-\infty}^{x}u_2(t,y)dy,\\ &\Omega(0,x) = \int_{-\infty}^{x}\omega(0,y)dy = \int_{-\infty}^{x}u_1(0,y)dy - \int_{-\infty}^{x}u_2(0,y)dy. \end{aligned} \tag{88}$$

is solution of the following Cauchy problem:

$$\begin{cases} \partial_t\omega + 3g\left(u_1^2\partial_x u_1 - u_2^2\partial_x u_2\right) & \\ \qquad -a\partial_x^3\omega + q\partial_x(u_1v_1 - u_2v_2) = b\Omega, & t>0,\ x\in\mathbb{R},\\ \partial_x\Omega = \omega, & t>0,\ x\in\mathbb{R},\\ \alpha\partial_x^2 V + \beta\partial_x V + \gamma V = \kappa(u_1^2 - u_2^2), & t>0,\ x\in\mathbb{R},\\ \Omega(t,-\infty) = 0, & t>0,\\ \omega(0,x) = u_{1,0} - u_{2,0}(x), & x\in\mathbb{R}. \end{cases} \tag{89}$$

Arguing as in ([15], Theorem 1.1), we have that

$$\|V(t,\cdot)\|^2_{H^2(\mathbb{R})} \le C(T)\,\|\omega(t,\cdot)\|^2_{L^2(\mathbb{R})}, \tag{90}$$

$$\|V(t,\cdot)\|^2_{L^\infty(\mathbb{R})} \le C(T)\,\|\omega(t,\cdot)\|^2_{L^2(\mathbb{R})}, \tag{91}$$

$$\|\partial_x V(t,\cdot)\|^2_{L^\infty(\mathbb{R})} \le C(T)\,\|\omega(t,\cdot)\|^2_{L^2(\mathbb{R})}. \tag{92}$$

Moreover, by (12) and (88),

$$\Omega(t,\infty) = \int_{\mathbb{R}} \omega(t,x)dx = \int_{\mathbb{R}} u_1(t,y)dy - \int_{\mathbb{R}} u_2(t,x)dx = 0. \tag{93}$$

Observe that, by (88)

$$\begin{aligned} 3g\left(u_1^2\partial_x u_1 - u_2^2\partial_x u_2\right) =& 3g\left(u_1^2\partial_x u_1 - u_2^2\partial_x u_1 + u_2^2\partial_x u_1 - u_2^2\partial_x u_2\right)\\ =& 3g\left(\partial_x u_1\left(u_1^2 - u_2^2\right) + u_2^2\partial_x\omega\right)\\ =& 3g\left(\partial_x u_1\left(u_1+u_2\right)\omega + u_2^2\partial_x\omega\right). \end{aligned} \tag{94}$$

Moreover, arguing as in ([15], Theorem 1.1),

$$q\partial_x(u_1v_1 - u_2v_2) = q\partial_x(u_1v_1 - u_2v_1 + u_2v_1 - u_2v_2) = q\partial_x(v_1\omega) + q\partial_x(u_2V). \tag{95}$$

Therefore, thanks to (94) and (95), the first equation of (89) is equivalent to the following one:

$$\partial_t\omega = b\Omega - 3g\partial_x u_1\left(u_1+u_2\right)\omega - 3gu_2^2\partial_x\omega + a\partial_x^3\omega - q\partial_x(v_1\omega) - q\partial_x(u_2V). \tag{96}$$

Multiplying (96) by $2\omega$, an integration on $\mathbb{R}$ gives

$$\begin{aligned} \frac{d}{dt}\|\omega(t,\cdot)\|^2_{L^2(\mathbb{R})} =& 2b\int_{\mathbb{R}}\Omega\partial_x\omega dx - 6g\int_{\mathbb{R}}\partial_x u_1\left(u_1+u_2\right)\omega^2 dx - 2a\int_{\mathbb{R}}\omega\partial_x^3\omega dx\\ &- 6g\int_{\mathbb{R}} u_2^2\omega\partial_x\omega dx - 2q\int_{\mathbb{R}}\partial_x(v_1\omega)\omega dx - 2q\int_{\mathbb{R}}\partial_x(u_2V)\omega dx. \end{aligned} \tag{97}$$

Observe that, by (88) and (93),

$$\begin{aligned} 2b\int_{\mathbb{R}}\Omega\partial_x\omega dx =& 2b\int_{\mathbb{R}}\Omega\partial_x\Omega dx = b\Omega^2(t,\infty) = 0,\\ -6g\int_{\mathbb{R}} u_2^2\omega\partial_x\omega dx =& 6g\int_{\mathbb{R}} u_2\partial_x u_2\omega^2 dx,\\ -2a\int_{\mathbb{R}}\omega\partial_x^3\omega dx =& 2a\int_{\mathbb{R}}\partial_x\omega\partial_x^2\omega = 0,\\ -2q\int_{\mathbb{R}}\partial_x(v_1\omega)\omega dx =& 2q\int_{\mathbb{R}} v_1\omega\partial_x\omega dx = -q\int_{\mathbb{R}}\partial_x v_1\omega^2 dx,\\ -2q\int_{\mathbb{R}}\partial_x(u_2V)\omega dx =& -2q\int_{\mathbb{R}}\partial_x u_2 V\omega dx - 2q\int_{\mathbb{R}} u_2\partial_x V\omega dx. \end{aligned} \tag{98}$$

It follows from (97) and (98) that

$$\begin{aligned} \frac{d}{dt}\|\omega(t,\cdot)\|^2_{L^2(\mathbb{R})} =& -6g\int_{\mathbb{R}}\partial_x u_1\left(u_1+u_2\right)\omega^2 dx + 6g\int_{\mathbb{R}} u_2\partial_x u_2\omega^2 dx\\ &- q\int_{\mathbb{R}}\partial_x v_1\omega^2 dx - 2q\int_{\mathbb{R}}\partial_x u_2 V\omega dx - 2q\int_{\mathbb{R}} u_2\partial_x V\omega dx. \end{aligned} \tag{99}$$

Since (11) holds, we have that

$$\begin{aligned}\|u_1\|_{L^\infty((0,T)\times\mathbb{R})}, \|u_2\|_{L^\infty((0,T)\times\mathbb{R})} \le& C(T),\\ \|\partial_x u_1\|_{L^\infty((0,T)\times\mathbb{R})}, \|\partial_x u_2\|_{L^\infty((0,T)\times\mathbb{R})} \le& C(T),\\ \|\partial_x v_1\|_{L^\infty((0,T)\times\mathbb{R})}, \|u_2(t,\cdot)\|_{L^2(\mathbb{R})}, \|\partial_x u_2(t,\cdot)\|_{L^2(\mathbb{R})} \le& C(T),\end{aligned} \tag{100}$$

for evert $0 \le t \le T$. Consequently, by (91), (100) and the Hölder inequality,

$$\begin{aligned}|6g|\int_{\mathbb{R}}|\partial_x u_1||u_1+u_2|\omega^2dx \le& |6g|\,\|\partial_x u_1\|_{L^\infty((0,T)\times\mathbb{R})}\int_{\mathbb{R}}|u_1+u_2|\omega^2dx\\ \le& C(T)\left(\|u_1\|_{L^\infty((0,T)\times\mathbb{R})}+\|u_2\|_{L^\infty((0,T)\times\mathbb{R})}\right)\|\omega(t,\cdot)\|^2_{L^2(\mathbb{R})}\\ \le& C(T)\,\|\omega(t,\cdot)\|^2_{L^2(\mathbb{R})},\\ |6g|\int_{\mathbb{R}}|u_2||\partial_x u_2|\omega^2dx \le& |6g|\,\|u_1\|_{L^\infty((0,T)\times\mathbb{R})}\int_{\mathbb{R}}|\partial_x u_2|\omega^2dx\\ \le& C(T)\,\|\partial_x u_2\|_{L^\infty((0,T)\times\mathbb{R})}\|\omega(t,\cdot)\|^2_{L^2(\mathbb{R})}\\ \le& C(T)\,\|\omega(t,\cdot)\|^2_{L^2(\mathbb{R})},\\ |q|\int_{\mathbb{R}}|\partial_x v_1|\omega^2dx \le& |q|\,\|\partial_x v_1\|_{L^\infty((0,T)\times\mathbb{R})}\|\omega(t,\cdot)\|^2_{L^2(\mathbb{R})}\\ \le& C(T)\,\|\omega(t,\cdot)\|^2_{L^2(\mathbb{R})},\\ |2q|\int_{\mathbb{R}}|\partial_x u_2||V||\omega|dx \le& |2q|\,\|V(t,\cdot)\|_{L^\infty(\mathbb{R})}\int_{\mathbb{R}}|\partial_x u_2||\omega|dx\\ \le& C(T)\,\|\partial_x u_2(t,\cdot)\|_{L^2(\mathbb{R})}\|\omega(t,\cdot)\|^2_{L^2(\mathbb{R})}\\ \le& C(T)\,\|\omega(t,\cdot)\|^2_{L^2(\mathbb{R})},\\ |2q|\int_{\mathbb{R}}|u_2||\partial_x V||\omega|dx \le& |2q|\,\|\partial_x V(t,\cdot)\|_{L^\infty(\mathbb{R})}\int_{\mathbb{R}}|u_2||\omega|dx\\ \le& C(T)\,\|u_2(t,\cdot)\|_{L^2(\mathbb{R})}\|\omega(t,\cdot)\|^2_{L^2(\mathbb{R})}\\ \le& C(T)\,\|\omega(t,\cdot)\|^2_{L^2(\mathbb{R})}.\end{aligned}$$

It follows from (99) that

$$\frac{d}{dt}\|\omega(t,\cdot)\|^2_{L^2(\mathbb{R})} \le C(T)\,\|\omega(t,\cdot)\|^2_{L^2(\mathbb{R})}. \tag{101}$$

The Gronwall Lemma and (89) give

$$\|\omega(t,\cdot)\|^2_{L^2(\mathbb{R})} \le e^{C(T)t}\,\|\omega(0,x)\|^2_{L^2(\mathbb{R})}. \tag{102}$$

Since (11) holds, by (88), arguing as in Lemma 5, $\Omega(t,\cdot)$ is integrable at $\pm\infty$. Moreover, thanks to (93) and Lemma 5, we have that

$$\int_{\mathbb{R}}\Omega(t,x)dx = 0. \tag{103}$$

Consider the following function:

$$\Omega_1(t,x) = \int_{-\infty}^{x}\Omega(t,y)dy, \tag{104}$$

since, by the second equation of (89),

$$\partial_t\Omega = \frac{d}{dt}\int_{-\infty}^{x}\omega(t,y)dy = \int_{-\infty}^{x}\partial_t\omega(t,y)dy, \tag{105}$$

integrating the first equation of (89) on $(-\infty, x)$, by (104) and (105), we have that

$$\partial_t \Omega = b\Omega_1 - g\left(u_1^3 - u_2^3\right) + a\partial_x^2 \omega - q\left(u_1 v_1 - u_2 v_2\right). \tag{106}$$

Observe that, by (88),

$$\begin{aligned} u_1^3 - u_2^3 &= \left(u_1^2 + u_2^2 + u_1 u_2\right)\omega, \\ u_1 v_1 - u_2 v_2 &= v_1 \omega + u_2 V. \end{aligned}$$

Consequently, by (106),

$$\partial_t \Omega = b\Omega_1 - g\left(u_1^2 + u_2^2 + u_1 u_2\right)\omega + a\partial_x^2 \omega - q v_1 \omega - q u_2 V. \tag{107}$$

It follows from (88), (93), (103) and (104) that

$$\begin{aligned} 2b\int_{\mathbb{R}} \Omega_1 \Omega dx &= 2b\int_{\mathbb{R}} \Omega_1 \partial_x \Omega_1 dx = b\Omega_1^2(t,\infty) = b\left(\int_{\mathbb{R}} \Omega(t,x)dx\right)^2 = 0, \\ 2a\int_{\mathbb{R}} \Omega \partial_x^2 \omega dx &= -2a\int_{\mathbb{R}} \partial_x \Omega \partial_x \omega dx = -2a\int_{\mathbb{R}} \omega \partial_x \omega dx = 0. \end{aligned}$$

Therefore, multiplying (107) by $2\Omega$, an integration on $\mathbb{R}$ gives

$$\begin{aligned} \frac{d}{dt}\|\Omega(t,\cdot)\|^2_{L^2(\mathbb{R})} = &-2g\int_{\mathbb{R}} \left(u_1^2 + u_2^2 + u_1 u_2\right)\omega \Omega dx \\ &- 2q\int_{\mathbb{R}} v_1 \omega \Omega dx - 2q\int_{\mathbb{R}} u_2 V \Omega dx. \end{aligned} \tag{108}$$

Due to (91), (100) and the Young inequality,

$$\begin{aligned} &|2g|\int_{\mathbb{R}} \left|u_1^2 + u_2^2 + u_1 u_2\right| |\omega||\Omega| dx \\ &\quad \le g^2 \int_{\mathbb{R}} \left(u_1^2 + u_2^2 + u_1 u_2\right)^2 \omega^2 dx + \|\Omega(t,\cdot)\|^2_{L^2(\mathbb{R})} \\ &\quad \le C(T)\|\omega(t,\cdot)\|^2_{L^2(\mathbb{R})} + \|\Omega(t,\cdot)\|^2_{L^2(\mathbb{R})}, \\ &|2q|\int_{\mathbb{R}} |v_1 \omega||\Omega| dx \\ &\quad \le q^2 \int_{\mathbb{R}} v_1^2 \omega^2 dx + \|\Omega(t,\cdot)\|^2_{L^2(\mathbb{R})} \\ &\quad \le q^2 \|v_1\|^2_{L^\infty((0,T)\times\mathbb{R})} \|\omega(t,\cdot)\|^2_{L^2(\mathbb{R})} + \|\Omega(t,\cdot)\|^2_{L^2(\mathbb{R})} \\ &\quad \le C(T)\|\omega(t,\cdot)\|^2_{L^2(\mathbb{R})} + \|\Omega(t,\cdot)\|^2_{L^2(\mathbb{R})}, \\ &|2q|\int_{\mathbb{R}} |u_2 V|\Omega dx \\ &\quad \le q^2 \int_{\mathbb{R}} V^2 u_2^2 dx + \|\Omega(t,\cdot)\|^2_{L^2(\mathbb{R})} \\ &\quad \le q^2 \|V(t,\cdot)\|^2_{L^\infty(\mathbb{R})} \|u_2(t,\cdot)\|^2_{L^2(\mathbb{R})} + \|\Omega(t,\cdot)\|^2_{L^2(\mathbb{R})} \\ &\quad \le C(T)\|V(t,\cdot)\|^2_{L^\infty(\mathbb{R})} + \|\Omega(t,\cdot)\|^2_{L^2(\mathbb{R})} \\ &\quad \le C(T)\|\omega(t,\cdot)\|^2_{L^2(\mathbb{R})} + \|\Omega(t,\cdot)\|^2_{L^2(\mathbb{R})}. \end{aligned}$$

Therefore, by (108),

$$\frac{d}{dt}\|\Omega(t,\cdot)\|^2_{L^2(\mathbb{R})} \le C(T)\|\omega(t,\cdot)\|^2_{L^2(\mathbb{R})} + 3\|\Omega(t,\cdot)\|^2_{L^2(\mathbb{R})}. \tag{109}$$

Adding (101) and (109), by (88) and the second equation of (89), we have that

$$\frac{d}{dt}\|\Omega(t,\cdot)\|^2_{H^1(\mathbb{R})} \le C(T)\|\omega(t,\cdot)\|^2_{L^2(\mathbb{R})} + 3\|\Omega(t,\cdot)\|^2_{L^2(\mathbb{R})} \le C(T)\|\Omega(t,\cdot)\|^2_{H^1(\mathbb{R})}$$

and

$$\|\Omega(t,\cdot)\|^2_{H^1(\mathbb{R})} \le e^{C(T)t}\|\Omega(0,\cdot)\|^2_{H^1(\mathbb{R})}. \tag{110}$$

Therefore, (13) follows (14), (88), (89), (90), (102) and (110). □

## 4. Conclusions

In this paper we studied the Cauchy problem for the Spectrum Pulse equation. It is a third order nonlocal nonlinear evolutive equation related to the dynamics of the electrical field of linearly polarized continuum spectrum pulses in optical waveguides. Our existence analysis is based on on passing to the limit in a fourth order perturbation of the equation. If the initial datum belongs to $H^2(\mathbb{R}) \cap L^1(\mathbb{R})$ and has zero mean we use the Aubin–Lions Lemma while if it belongs to $H^3(\mathbb{R}) \cap L^1(\mathbb{R})$ and has zero mean we use the Sobolev Immersion Theorem. Finally, we directly prove a stability estimate that implies the uniqueness of the solution.

**Author Contributions:** Writing—original draft, G.M.C. and L.d.R.

**Funding:** This research received no external funding.

**Acknowledgments:** The authors are members of the Gruppo Nazionale per l'Analisi Matematica, la Probabilità e le loro Applicazioni (GNAMPA) of the Istituto Nazionale di Alta Matematica (INdAM).

**Conflicts of Interest:** The authors declare no conflict of interest.

## Appendix A. $u_0 \in H^3(\mathbb{R}) \cap L^1(\mathbb{R})$

In this appendix, we consider the Cauchy problem (1), where, on the initial datum, we assume

$$u_0(x) \in H^3(\mathbb{R}) \cap L^1(\mathbb{R}), \quad \int_{\mathbb{R}} u_0(x)dx = 0, \tag{A1}$$

while on the function $P(x)$, defined in (3), we assume (4). Moreover, we assume (5). The main result of this appendix is the following theorem.

**Theorem A1.** *Assume (3), (4), (5) and (A1). Fix $T > 0$, there exists an unique solution $(u, v, P)$ of (1) such that*

$$\begin{aligned}
&u \in H^1((0,T)\times\mathbb{R}) \cap L^\infty(0,T;H^3(\mathbb{R})),\\
&v \in H^1((0,T)\times\mathbb{R}) \cap L^\infty(0,T;H^5(\mathbb{R})) \cap W^{1,\infty}((0,T)\times\mathbb{R}),\\
&\partial^2_{tx}v \in L^\infty((0,T)\times\mathbb{R}) \cap L^\infty(0,T;L^2(\mathbb{R})),\\
&P \in L^\infty(0,T;H^4(\mathbb{R})).
\end{aligned} \tag{A2}$$

*Moreover, (12) and (13) hold.*

To prove Theorem A1, we consider the approximation (16), where $u_{\varepsilon,0}$ is a $C^\infty$ approximation of $u_0$ such that

$$\begin{aligned}
&\|u_{\varepsilon,0}\|_{H^3(\mathbb{R})} \le \|u_0\|_{H^3(\mathbb{R})}, \quad \int_{\mathbb{R}} u_{\varepsilon,0}dx = 0,\\
&\|P_{\varepsilon,0}\|_{L^2(\mathbb{R})} \le \|P_0\|_{L^2(\mathbb{R})}, \quad \int_{\mathbb{R}} P_{\varepsilon,0}dx = 0,\\
&\varepsilon\left\|\partial_x^4 u_\varepsilon(t,\cdot)\right\|^2_{L^2(\mathbb{R})} \le C_0,
\end{aligned} \tag{A3}$$

where $C_0$ is a positive constant, independent on $\varepsilon$.

Let us prove some a priori estimates on $u_\varepsilon$, $v_\varepsilon$ and $P_\varepsilon$.

Since $H^2(\mathbb{R}) \subset H^3(\mathbb{R})$, then Lemmas 1, 2, 3, 4, 5, 6, 7, 8, 9 and 10 are still valid.

We prove the following result.

**Lemma A1.** *Assume (5). Fix $T > 0$. There exists a constant $C(T) > 0$, independent on $\varepsilon$, such that,*

$$\left\|\partial_x^3 u_\varepsilon\right\|_{L^\infty((0,T)\times\mathbb{R})} \leq C(T). \tag{A4}$$

*In particular, we have that*

$$\left\|\partial_x^3 u_\varepsilon(t,\cdot)\right\|^2_{L^2(\mathbb{R})} + 2\varepsilon e^{C(T)t}\int_0^t e^{-C(T)s}\left\|\partial_x^5 u_\varepsilon(s,\cdot)\right\|^2_{L^2(\mathbb{R})} ds \leq C(T), \tag{A5}$$

*for every $0 \leq t \leq T$.*

**Proof.** Let $0 \leq t \leq T$. Multiplying the first equation of (16) by $-2\partial_x^6 u_\varepsilon$, we have that

$$\begin{aligned} -2\partial_x^6 u_\varepsilon \partial_t u_\varepsilon =& -2bP_\varepsilon\partial_x^6 u_\varepsilon + 2\varepsilon\partial_x^6 u_\varepsilon\partial_x^4 u_\varepsilon + 6gu_\varepsilon^2\partial_x u_\varepsilon\partial_x^6 u_\varepsilon \\ &- 2a\partial_x^3 u_\varepsilon\partial_x^6 u_\varepsilon + 2qu_\varepsilon\partial_x v_\varepsilon\partial_x^6 u_\varepsilon + 2qv_\varepsilon\partial_x u_\varepsilon\partial_x^6 u_\varepsilon. \end{aligned} \tag{A6}$$

Observe that by (18) and the second equation of (16),

$$\begin{aligned} -2b\int_{\mathbb{R}} P_\varepsilon\partial_x^6 u_\varepsilon dx =& 2b\int_{\mathbb{R}}\partial_x P_\varepsilon\partial_x^5 u_\varepsilon dx = 2b\int_{\mathbb{R}} u_\varepsilon\partial_x^5 u_\varepsilon dx \\ =& -2b\int_{\mathbb{R}}\partial_x u_\varepsilon\partial_x^4 u_\varepsilon dx = 2b\int_{\mathbb{R}}\partial_x^2 u_\varepsilon\partial_x^3 u_\varepsilon dx = 0. \end{aligned} \tag{A7}$$

Moreover,

$$\begin{aligned} -2\int_{\mathbb{R}}\partial_x^6 u_\varepsilon\partial_t u_\varepsilon =& \frac{d}{dt}\left\|\partial_x^3 u_\varepsilon(t,\cdot)\right\|^2_{L^2(\mathbb{R})}, \\ 2\varepsilon\int_{\mathbb{R}}\partial_x^6 u_\varepsilon\partial_x^4 u_\varepsilon dx =& -2\varepsilon\left\|\partial_x^5 u_\varepsilon(t,\cdot)\right\|^2_{L^2(\mathbb{R})}, \\ -2a\int_{\mathbb{R}}\partial_x^3 u_\varepsilon\partial_x^6 u_\varepsilon dx =& 2a\int_{\mathbb{R}}\partial_x^4 u_\varepsilon\partial_x^5 u_\varepsilon dx = 0. \end{aligned} \tag{A8}$$

It follows from (A7), (A8) and an integration of (A6) on $\mathbb{R}$ that

$$\begin{aligned} \frac{d}{dt}&\left\|\partial_x^3 u_\varepsilon(t,\cdot)\right\|^2_{L^2(\mathbb{R})} + 2\varepsilon\left\|\partial_x^5 u_\varepsilon(t,\cdot)\right\|^2_{L^2(\mathbb{R})} \\ &= 6g\int_{\mathbb{R}} u_\varepsilon^2\partial_x u_\varepsilon\partial_x^6 u_\varepsilon dx + 2q\int_{\mathbb{R}} u_\varepsilon\partial_x v_\varepsilon\partial_x^6 u_\varepsilon dx \\ &\quad + 2q\int_{\mathbb{R}} v_\varepsilon\partial_x u_\varepsilon\partial_x^6 u_\varepsilon dx. \end{aligned} \tag{A9}$$

Observe that

$$\begin{aligned}
6g\int_{\mathbb{R}} u_\varepsilon^2\partial_x u_\varepsilon\partial_x^6 u_\varepsilon dx =& -12g\int_{\mathbb{R}} u_\varepsilon(\partial_x u_\varepsilon)^2\partial_x^5 u_\varepsilon dx - 6g\int_{\mathbb{R}} u_\varepsilon^2\partial_x^2 u_\varepsilon\partial_x^5 u_\varepsilon dx\\
=&12g\int_{\mathbb{R}}(\partial_x u_\varepsilon)^3\partial_x^4 u_\varepsilon dx + 36g\int_{\mathbb{R}} u_\varepsilon\partial_x u_\varepsilon\partial_x^2 u_\varepsilon\partial_x^4 u_\varepsilon dx\\
&+6g\int_{\mathbb{R}} u_\varepsilon^2\partial_x^3 u_\varepsilon\partial_x^4 u_\varepsilon dx\\
=&-72g\int_{\mathbb{R}}(\partial_x u_\varepsilon)^2\partial_x^2 u_\varepsilon\partial_x^3 u_\varepsilon dx - 36g\int_{\mathbb{R}} u_\varepsilon(\partial_x^2 u_\varepsilon)^2\partial_x^3 u_\varepsilon dx\\
&-42g\int_{\mathbb{R}} u_\varepsilon\partial_x u_\varepsilon(\partial_x^3 u_\varepsilon)^2 dx,\\
2q\int_{\mathbb{R}} u_\varepsilon\partial_x v_\varepsilon\partial_x^6 u_\varepsilon dx =& -2q\int_{\mathbb{R}}\partial_x u_\varepsilon\partial_x v_\varepsilon\partial_x^5 u_\varepsilon dx - 2q\int_{\mathbb{R}} u_\varepsilon\partial_x^2 v_\varepsilon\partial_x^5 u_\varepsilon dx\\
=&2q\int_{\mathbb{R}}\partial_x^2 u_\varepsilon\partial_x v_\varepsilon\partial_x^4 u_\varepsilon dx + 4q\int_{\mathbb{R}}\partial_x u_\varepsilon\partial_x^2 v_\varepsilon\partial_x^4 u_\varepsilon dx\\
&+2q\int_{\mathbb{R}} u_\varepsilon\partial_x^3 v_\varepsilon\partial_x^4 u_\varepsilon dx\\
=&-2q\int_{\mathbb{R}}\partial_x v_\varepsilon(\partial_x^3 u_\varepsilon)^2 dx - 6q\int_{\mathbb{R}}\partial_x^2 u_\varepsilon\partial_x^2 v_\varepsilon\partial_x^3 u_\varepsilon dx\\
&-6q\int_{\mathbb{R}}\partial_x u_\varepsilon\partial_x^3 v_\varepsilon\partial_x^3 u_\varepsilon dx - 2q\int_{\mathbb{R}} u_\varepsilon\partial_x^4 v_\varepsilon\partial_x^3 u_\varepsilon dx\\
=&-2q\int_{\mathbb{R}}\partial_x v_\varepsilon(\partial_x^3 u_\varepsilon)^2 dx + 3q\int_{\mathbb{R}}\partial_x^3 v_\varepsilon(\partial_x^2 u_\varepsilon)^2 dx\\
&-6q\int_{\mathbb{R}}\partial_x u_\varepsilon\partial_x^3 v_\varepsilon\partial_x^3 u_\varepsilon dx - 2q\int_{\mathbb{R}} u_\varepsilon\partial_x^4 v_\varepsilon\partial_x^3 u_\varepsilon dx,\\
2q\int_{\mathbb{R}} v_\varepsilon\partial_x u_\varepsilon\partial_x^6 u_\varepsilon dx =& -2q\int_{\mathbb{R}}\partial_x v_\varepsilon\partial_x u_\varepsilon\partial_x^5 u_\varepsilon dx - 2q\int_{\mathbb{R}} v_\varepsilon\partial_x^2 u_\varepsilon\partial_x^5 u_\varepsilon dx\\
=&2q\int_{\mathbb{R}}\partial_x^2 v_\varepsilon\partial_x u_\varepsilon\partial_x^4 u_\varepsilon dx + 4q\int_{\mathbb{R}}\partial_x v_\varepsilon\partial_x^2 u_\varepsilon\partial_x^4 u_\varepsilon dx\\
&+2q\int_{\mathbb{R}} v_\varepsilon\partial_x^3 u_\varepsilon\partial_x^4 u_\varepsilon dx\\
=&-2q\int_{\mathbb{R}}\partial_x^3 v_\varepsilon\partial_x u_\varepsilon\partial_x^3 u_\varepsilon dx - 6q\int_{\mathbb{R}}\partial_x^2 v_\varepsilon\partial_x^2 u_\varepsilon\partial_x^3 u_\varepsilon dx\\
&-5q\int_{\mathbb{R}}\partial_x v_\varepsilon(\partial_x^3 u_\varepsilon)^2 dx\\
=&-2q\int_{\mathbb{R}}\partial_x^3 v_\varepsilon\partial_x u_\varepsilon\partial_x^3 u_\varepsilon dx + 3q\int_{\mathbb{R}}\partial_x^3 v_\varepsilon(\partial_x^2 u_\varepsilon)^2 dx\\
&-5q\int_{\mathbb{R}}\partial_x v_\varepsilon(\partial_x^3 u_\varepsilon)^2 dx.
\end{aligned}$$

Consequently, by (A9),

$$\begin{aligned}
\frac{d}{dt}&\left\|\partial_x^3 u_\varepsilon(t,\cdot)\right\|^2_{L^2(\mathbb{R})} + 2\varepsilon\left\|\partial_x^5 u_\varepsilon(t,\cdot)\right\|^2_{L^2(\mathbb{R})}\\
=&-72g\int_{\mathbb{R}}(\partial_x u_\varepsilon)^2\partial_x^2 u_\varepsilon\partial_x^3 u_\varepsilon dx - 36g\int_{\mathbb{R}} u_\varepsilon(\partial_x^2 u_\varepsilon)^2\partial_x^3 u_\varepsilon dx\\
&-42g\int_{\mathbb{R}} u_\varepsilon\partial_x u_\varepsilon(\partial_x^3 u_\varepsilon)^2 dx - 7q\int_{\mathbb{R}}\partial_x v_\varepsilon(\partial_x^3 u_\varepsilon)^2 dx\\
&+6q\int_{\mathbb{R}}\partial_x^3 v_\varepsilon(\partial_x^2 u_\varepsilon)^2 dx - 8q\int_{\mathbb{R}}\partial_x^3 v_\varepsilon\partial_x u_\varepsilon\partial_x^3 u_\varepsilon dx\\
&-2q\int_{\mathbb{R}} u_\varepsilon\partial_x^4 v_\varepsilon\partial_x^3 u_\varepsilon dx.
\end{aligned}\tag{A10}$$

Due to (24), (41), (42), (56), (57), (68), (69) and the Young inequality,

$$\begin{aligned}
&|72g|\int_{\mathbb{R}}|(\partial_x u_\varepsilon)^2\partial_x^2 u_\varepsilon||\partial_x^3 u_\varepsilon|dx\\
&\quad\le 36g^2\int_{\mathbb{R}}(\partial_x u_\varepsilon)^4(\partial_x^2 u_\varepsilon)^2dx+36\left\|\partial_x^3 u_\varepsilon(t,\cdot)\right\|^2_{L^2(\mathbb{R})}\\
&\quad\le 36g^2\left\|\partial_x u_\varepsilon\right\|^4_{L^\infty((0,T)\times\mathbb{R})}\left\|\partial_x^2 u_\varepsilon(t,\cdot)\right\|^2_{L^2(\mathbb{R})}+36\left\|\partial_x^3 u_\varepsilon(t,\cdot)\right\|^2_{L^2(\mathbb{R})}\\
&\quad\le C(T)+36\left\|\partial_x^3 u_\varepsilon(t,\cdot)\right\|^2_{L^2(\mathbb{R})},\\
&|36g|\int_{\mathbb{R}}|u_\varepsilon(\partial_x^2 u_\varepsilon)^2||\partial_x^3 u_\varepsilon|dx\\
&\quad\le 18g^2\int_{\mathbb{R}}u_\varepsilon^2(\partial_x^2 u_\varepsilon)^4dx+18\left\|\partial_x^3 u_\varepsilon(t,\cdot)\right\|^2_{L^2(\mathbb{R})}\\
&\quad\le 18g^2\left\|u_\varepsilon\right\|^2_{L^\infty((0,T)\times\mathbb{R})}\left\|\partial_x^2 u_\varepsilon\right\|^2_{L^\infty((0,T)\times\mathbb{R})}\left\|\partial_x^2 u_\varepsilon(t,\cdot)\right\|^2_{L^2(\mathbb{R})}+18\left\|\partial_x^3 u_\varepsilon(t,\cdot)\right\|^2_{L^2(\mathbb{R})}\\
&\quad\le C(T)\left\|\partial_x^2 u_\varepsilon\right\|^2_{L^\infty((0,T)\times\mathbb{R})}+18\left\|\partial_x^3 u_\varepsilon(t,\cdot)\right\|^2_{L^2(\mathbb{R})},\\
&|7q|\int_{\mathbb{R}}|\partial_x v_\varepsilon|(\partial_x^3 u_\varepsilon)^2dx\le|7q|\left\|\partial_x v_\varepsilon(t,\cdot)\right\|_{L^\infty(\mathbb{R})}\left\|\partial_x^3 u_\varepsilon(t,\cdot)\right\|^2_{L^2(\mathbb{R})}\\
&\quad\le C_0\left\|\partial_x^3 u_\varepsilon(t,\cdot)\right\|^2_{L^2(\mathbb{R})},\\
&|6q|\int_{\mathbb{R}}|\partial_x^3 v_\varepsilon|(\partial_x^2 u_\varepsilon)^2dx\\
&\quad\le|6q|\left\|\partial_x^3 v_\varepsilon\right\|_{L^\infty((0,T)\times\mathbb{R})}\left\|\partial_x^2 u_\varepsilon(t,\cdot)\right\|^2_{L^2(\mathbb{R})}\le C(T),\\
&|8q|\int_{\mathbb{R}}|\partial_x^3 v_\varepsilon\partial_x u_\varepsilon||\partial_x^3 u_\varepsilon|dx\\
&\quad\le 4q^2\int_{\mathbb{R}}(\partial_x^3 v_\varepsilon)^2(\partial_x u_\varepsilon)^2dx+4\left\|\partial_x^3 u_\varepsilon(t,\cdot)\right\|^2_{L^2(\mathbb{R})}\\
&\quad\le 4q^2\left\|\partial_x^3 v_\varepsilon\right\|^2_{L^\infty((0,T)\times\mathbb{R})}\left\|\partial_x u_\varepsilon(t,\cdot)\right\|^2_{L^2(\mathbb{R})}+4\left\|\partial_x^3 u_\varepsilon(t,\cdot)\right\|^2_{L^2(\mathbb{R})}\\
&\quad\le C(T)+4\left\|\partial_x^3 u_\varepsilon(t,\cdot)\right\|^2_{L^2(\mathbb{R})},\\
&|2q|\int_{\mathbb{R}}|u_\varepsilon\partial_x^4 v_\varepsilon||\partial_x^3 u_\varepsilon|dx\\
&\quad\le q^2\int_{\mathbb{R}}u_\varepsilon^2(\partial_x^4 v_\varepsilon)^2dx+\left\|\partial_x^3 u_\varepsilon(t,\cdot)\right\|^2_{L^2(\mathbb{R})},\\
&\quad\le q^2\left\|u_\varepsilon\right\|^2_{L^\infty((0,T)\times\mathbb{R})}\left\|\partial_x^4 v_\varepsilon(t,\cdot)\right\|^2_{L^2(\mathbb{R})}+\left\|\partial_x^3 u_\varepsilon(t,\cdot)\right\|^2_{L^2(\mathbb{R})}.
\end{aligned}$$

It follows from (A10) that

$$\begin{aligned}
&\frac{d}{dt}\left\|\partial_x^3 u_\varepsilon(t,\cdot)\right\|^2_{L^2(\mathbb{R})}+2\varepsilon\left\|\partial_x^5 u_\varepsilon(t,\cdot)\right\|^2_{L^2(\mathbb{R})}\\
&\qquad\le C(T)\left\|\partial_x^3 u_\varepsilon(t,\cdot)\right\|^2_{L^2(\mathbb{R})}+C(T)\left(1+\left\|\partial_x^2 u_\varepsilon\right\|^2_{L^\infty((0,T)\times\mathbb{R})}\right).
\end{aligned}$$

The Gronwall Lemma and (A3) give

$$\begin{aligned}\left\|\partial_x^3 u_\varepsilon(t,\cdot)\right\|^2_{L^2(\mathbb{R})} &+ 2\varepsilon e^{C(T)t}\int_0^t e^{-C(T)s}\left\|\partial_x^5 u_\varepsilon(s,\cdot)\right\|^2_{L^2(\mathbb{R})} ds\\ &\le C_0 e^{C(T)t} + C(T)\left(1+\left\|\partial_x^2 u_\varepsilon\right\|^2_{L^\infty((0,T)\times\mathbb{R})}\right)e^{C(T)t}\int_0^t e^{-C(T)s}ds \\ &\le C(T)\left(1+\left\|\partial_x^2 u_\varepsilon\right\|^2_{L^\infty((0,T)\times\mathbb{R})}\right).\end{aligned} \tag{A11}$$

We prove (A4). Thanks to (57), (A11) and the Hölder inequality,

$$\begin{aligned}(\partial_x^2 u_\varepsilon(t,x))^2 =&2\int_{-\infty}^x \partial_x^2 u_\varepsilon \partial_x^3 u_\varepsilon dx \le 2\int_{\mathbb{R}} |\partial_x^2 u_\varepsilon||\partial_x^3 u_\varepsilon| dx\\ &\le \left\|\partial_x^2 u_\varepsilon(t,\cdot)\right\|_{L^2(\mathbb{R})}\left\|\partial_x^3 u_\varepsilon(t,\cdot)\right\|^2_{L^2(\mathbb{R})} \le C(T)\sqrt{\left(1+\|\partial_x^2 u_\varepsilon\|^2_{L^\infty((0,T)\times\mathbb{R})}\right)}.\end{aligned}$$

Hence,

$$\left\|\partial_x^2 u_\varepsilon\right\|^4_{L^\infty((0,T)\times\mathbb{R})} - C(T)\left\|\partial_x^2 u_\varepsilon\right\|^2_{L^\infty((0,T)\times\mathbb{R})} - C(T) \le 0,$$

which gives (A4).

Finally, (A5) follows from (A4) and (A11). □

**Lemma A2.** *Assume (5). Fix $T > 0$. There exists a constant $C(T) > 0$, independent on $\varepsilon$, such that*

$$\left\|\partial_x^5 v_\varepsilon(t,\cdot)\right\|^2_{L^2(\mathbb{R})} \le C(T), \tag{A12}$$

*for every $0 \le t \le T$. In particular, we have that*

$$\left\|\partial_x^4 v_\varepsilon\right\|_{L^\infty((0,T)\times\mathbb{R})} \le C(T). \tag{A13}$$

**Proof.** Let $0 \le t \le T$. Differentiating (70) with respect to $x$, we have

$$\alpha\partial_x^5 v_\varepsilon = 6\kappa\partial_x u_\varepsilon\partial_x^2 u_\varepsilon + 2\kappa u_\varepsilon\partial_x^3 u_\varepsilon - \beta\partial_x^4 v_\varepsilon - \gamma\partial_x^3 v_\varepsilon. \tag{A14}$$

Since $\partial_x^3 u_\varepsilon(t,\pm\infty) = 0$, by (55), (71) and (72), we have that

$$\partial_x^5 v_\varepsilon(t,\pm\infty) = 0. \tag{A15}$$

Observe that

$$\begin{aligned}-2\beta\alpha\int_{\mathbb{R}}\partial_x^4 v_\varepsilon\partial_x^5 v_\varepsilon dx =&0,\\ -2\alpha\gamma\int_{\mathbb{R}}\partial_x^3 v_\varepsilon\partial_x^5 v_\varepsilon dx =&2\alpha\gamma\left\|\partial_x^4 v_\varepsilon(t,\cdot)\right\|^2_{L^2(\mathbb{R})}.\end{aligned}$$

Consequently, multiplying (A14) by $2\alpha\partial_x^5 v_\varepsilon$, an integration on $\mathbb{R}$ gives

$$\begin{aligned}2\alpha^2\left\|\partial_x^5 v_\varepsilon(t,\cdot)\right\|^2_{L^2(\mathbb{R})} =&12\alpha\kappa\int_{\mathbb{R}}\partial_x u_\varepsilon\partial_x^2 u_\varepsilon\partial_x^5 v_\varepsilon dx + 4\alpha\kappa\int_{\mathbb{R}} u_\varepsilon\partial_x^3 u_\varepsilon\partial_x^5 v_\varepsilon dx\\ &+ 2\alpha\gamma\left\|\partial_x^4 v_\varepsilon(t,\cdot)\right\|^2_{L^2(\mathbb{R})}.\end{aligned} \tag{A16}$$

Due to (41), (56), (57), (A4) and the Young inequality,

$$
\begin{aligned}
&12|\alpha\kappa|\int_{\mathbb{R}}|\partial_x u_\varepsilon\partial_x^2 u_\varepsilon||\partial_x^5 v_\varepsilon|dx\\
&=\int_{\mathbb{R}}|12\kappa\partial_x u_\varepsilon\partial_x^2 u_\varepsilon||\alpha\partial_x^5 v_\varepsilon|dx\\
&\le 72\kappa^2\int_{\mathbb{R}}(\partial_x u_\varepsilon)^2(\partial_x^2 u_\varepsilon)^2dx+\frac{\alpha^2}{2}\left\|\partial_x^5 v_\varepsilon(t,\cdot)\right\|^2_{L^2(\mathbb{R})}\\
&\le 72\kappa^2\left\|\partial_x u_\varepsilon\right\|^2_{L^\infty((0,T)\times\mathbb{R})}\left\|\partial_x^2 u_\varepsilon(t,\cdot)\right\|^2_{L^2(\mathbb{R})}+\frac{\alpha^2}{2}\left\|\partial_x^5 v_\varepsilon(t,\cdot)\right\|^2_{L^2(\mathbb{R})}\\
&\le C(T)+\frac{\alpha^2}{2}\left\|\partial_x^5 v_\varepsilon(t,\cdot)\right\|^2_{L^2(\mathbb{R})},\\
&|4\alpha\kappa|\int_{\mathbb{R}}|u_\varepsilon\partial_x^3 u_\varepsilon||\partial_x^5 v_\varepsilon|dx\\
&=\int_{\mathbb{R}}|4\kappa u_\varepsilon\partial_x^3 u_\varepsilon|||\alpha\partial_x^5 v_\varepsilon|dx\\
&\le 8\kappa^2\int_{\mathbb{R}}u_\varepsilon^2(\partial_x^3 u_\varepsilon)^2dx+\frac{\alpha^2}{2}\left\|\partial_x^5 v_\varepsilon(t,\cdot)\right\|^2_{L^2(\mathbb{R})}\\
&\le 8\kappa^2\left\|u_\varepsilon\right\|^2_{L^\infty((0,T)\times\mathbb{R})}\left\|\partial_x^3 u_\varepsilon(t,\cdot)\right\|^2_{L^2(\mathbb{R})}+\frac{\alpha^2}{2}\left\|\partial_x^5 v_\varepsilon(t,\cdot)\right\|^2_{L^2(\mathbb{R})}\\
&\le C(T)+\frac{\alpha^2}{2}\left\|\partial_x^5 v_\varepsilon(t,\cdot)\right\|^2_{L^2(\mathbb{R})}.
\end{aligned}
$$

It follows from (68) and (A16) that

$$
\alpha^2\left\|\partial_x^5 v_\varepsilon(t,\cdot)\right\|^2_{L^2(\mathbb{R})}\le C(T)+|2\alpha\gamma|\left\|\partial_x^4 v_\varepsilon(t,\cdot)\right\|^2_{L^2(\mathbb{R})}\le C(T),
$$

which gives (A12).

Finally, we prove (A13). Thanks to (68), (A12) and the Hölder inequality,

$$
\begin{aligned}
(\partial_x^4 v_\varepsilon(t,x))^2=&2\int_{-\infty}^{x}\partial_x^4 v_\varepsilon\partial_x^5 v_\varepsilon dx\le 2\int_{\mathbb{R}}|\partial_x^4 v_\varepsilon||\partial_x^5 v_\varepsilon|dx\\
&\le\left\|\partial_x^4 v_\varepsilon(t,\cdot)\right\|_{L^2(\mathbb{R})}\left\|\partial_x^5 v_\varepsilon(t,\cdot)\right\|_{L^2(\mathbb{R})}\le C(T).
\end{aligned}
$$

Hence,

$$
\left\|\partial_x^4 v_\varepsilon\right\|^2_{L^\infty((0,T)\times\mathbb{R})}\le C(T),
$$

which gives (A13). □

**Lemma A3.** *Assume* (5). *Fix $T>0$. There exists a constant $C(T)>0$, independent on $\varepsilon$, such that,*

$$
\varepsilon\left\|\partial_x^4 u_\varepsilon(t,\cdot)\right\|^2_{L^2(\mathbb{R})}+\varepsilon^2\int_0^t\left\|\partial_x^6 u_\varepsilon(s,\cdot)\right\|^2_{L^2(\mathbb{R})}ds\le C(T), \tag{A17}
$$

*for every $0\le t\le T$.*

**Proof.** Let $0\le t\le T$. Multiplying the first equation of (16) by $2\varepsilon\partial_x^8 u_\varepsilon$, we have

$$
\begin{aligned}
2\varepsilon\partial_x^8 u_\varepsilon\partial_t u_\varepsilon=&2b\varepsilon P_\varepsilon\partial_x^8 u_\varepsilon-2\varepsilon^2\partial_x^4 u_\varepsilon\partial_x^8 u_\varepsilon-6g\varepsilon u_\varepsilon^2\partial_x u_\varepsilon\partial_x^8 u_\varepsilon\\
&+2a\varepsilon\partial_x^3 u_\varepsilon\partial_x^8 u_\varepsilon-2q\varepsilon u_\varepsilon\partial_x v_\varepsilon\partial_x^8 u_\varepsilon-2q\varepsilon v_\varepsilon\partial_x u_\varepsilon\partial_x^8 u_\varepsilon.
\end{aligned} \tag{A18}
$$

Observe that by the second equation of (16) and (18),

$$\begin{aligned}2b\varepsilon\int_{\mathbb{R}} P_\varepsilon\partial_x^8 u_\varepsilon dx =&-2b\varepsilon\int_{\mathbb{R}}\partial_x P_\varepsilon\partial_x^7 u_\varepsilon dx=-2b\varepsilon\int_{\mathbb{R}} u_\varepsilon\partial_x^7 u_\varepsilon dx=2b\varepsilon\int_{\mathbb{R}}\partial_x u_\varepsilon\partial_x^6 u_\varepsilon dx\\ =&-2b\varepsilon\int_{\mathbb{R}}\partial_x^2 u_\varepsilon\partial_x^5 u_\varepsilon dx=2b\varepsilon\int_{\mathbb{R}}\partial_x^3 u_\varepsilon\partial_x^4 u_\varepsilon dx=0.\end{aligned}\tag{A19}$$

Moreover,

$$\begin{aligned}2\varepsilon\int_{\mathbb{R}}\partial_x^8 u_\varepsilon\partial_t u_\varepsilon dx =&\varepsilon\frac{d}{dt}\left\|\partial_x^4 u_\varepsilon(t,\cdot)\right\|^2_{L^2(\mathbb{R})},\\ -2\varepsilon^2\int_{\mathbb{R}}\partial_x^4 u_\varepsilon\partial_x^8 u_\varepsilon dx =&-2\varepsilon^2\left\|\partial_x^6 u_\varepsilon(t,\cdot)\right\|^2_{L^2(\mathbb{R})},\\ 2a\varepsilon\int_{\mathbb{R}}\partial_x^3 u_\varepsilon\partial_x^8 u_\varepsilon dx =&0.\end{aligned}\tag{A20}$$

Therefore, (A19), (A20) and an integration of (A18) on $\mathbb{R}$ give

$$\begin{aligned}\varepsilon\frac{d}{dt}\left\|\partial_x^4 u_\varepsilon(t,\cdot)\right\|^2_{L^2(\mathbb{R})}&+2\varepsilon^2\left\|\partial_x^6 u_\varepsilon(t,\cdot)\right\|^2_{L^2(\mathbb{R})}\\ &=-6g\varepsilon\int_{\mathbb{R}} u_\varepsilon^2\partial_x u_\varepsilon\partial_x^8 u_\varepsilon dx-2q\varepsilon\int_{\mathbb{R}} u_\varepsilon\partial_x v_\varepsilon\partial_x^8 u_\varepsilon dx\\ &\quad-2q\varepsilon\int_{\mathbb{R}} v_\varepsilon\partial_x u_\varepsilon\partial_x^8 u_\varepsilon dx.\end{aligned}\tag{A21}$$

Observe that

$$\begin{aligned}-6g\varepsilon\int_{\mathbb{R}} u_\varepsilon^2\partial_x u_\varepsilon\partial_x^8 u_\varepsilon dx =&12g\varepsilon\int u_\varepsilon(\partial_x u_\varepsilon)^2\partial_x^7 u_\varepsilon dx+6g\varepsilon\int_{\mathbb{R}} u_\varepsilon^2\partial_x^2 u_\varepsilon\partial_x^7 u_\varepsilon dx\\ =&-12g\varepsilon\int_{\mathbb{R}}(\partial_x u_\varepsilon)^3\partial_x^6 u_\varepsilon dx-36g\varepsilon\int_{\mathbb{R}} u_\varepsilon\partial_x u_\varepsilon\partial_x^2 u_\varepsilon\partial_x^6 u_\varepsilon dx\\ &-6g\varepsilon\int_{\mathbb{R}} u_\varepsilon^2\partial_x^3 u_\varepsilon\partial_x^6 u_\varepsilon dx,\\ -2q\varepsilon\int_{\mathbb{R}} u_\varepsilon\partial_x v_\varepsilon\partial_x^8 u_\varepsilon dx =&2q\varepsilon\int_{\mathbb{R}}\partial_x u_\varepsilon\partial_x v_\varepsilon\partial_x^7 u_\varepsilon dx+2q\varepsilon\int_{\mathbb{R}} u_\varepsilon\partial_x^2 v_\varepsilon\partial_x^7 u_\varepsilon dx\\ =&-2q\varepsilon\int_{\mathbb{R}}\partial_x^2 u_\varepsilon\partial_x v_\varepsilon\partial_x^6 u_\varepsilon dx-4q\varepsilon\int_{\mathbb{R}}\partial_x u_\varepsilon\partial_x^2 v_\varepsilon\partial_x^6 u_\varepsilon dx\\ &-2q\varepsilon\int_{\mathbb{R}} u_\varepsilon\partial_x^3 v_\varepsilon\partial_x^6 u_\varepsilon dx,\\ -2q\varepsilon\int_{\mathbb{R}} v_\varepsilon\partial_x u_\varepsilon\partial_x^8 u_\varepsilon dx =&2q\varepsilon\int_{\mathbb{R}}\partial_x v_\varepsilon\partial_x u_\varepsilon\partial_x^6 u_\varepsilon dx+2q\varepsilon\int_{\mathbb{R}} v_\varepsilon\partial_x^2 u_\varepsilon\partial_x^7 u_\varepsilon dx\\ =&-2q\varepsilon\int_{\mathbb{R}}\partial_x^2 v_\varepsilon\partial_x u_\varepsilon\partial_x^6 u_\varepsilon dx-4q\varepsilon\int_{\mathbb{R}}\partial_x v_\varepsilon\partial_x^2 u_\varepsilon\partial_x^6 u_\varepsilon dx\\ &-2q\varepsilon\int_{\mathbb{R}} v_\varepsilon\partial_x^3 u_\varepsilon\partial_x^6 u_\varepsilon dx.\end{aligned}\tag{A22}$$

Consequently, by (A21),

$$\begin{aligned}
\varepsilon\frac{d}{dt}\left\|\partial_x^4 u_\varepsilon(t,\cdot)\right\|^2_{L^2(\mathbb{R})}&+2\varepsilon^2\left\|\partial_x^6 u_\varepsilon(t,\cdot)\right\|^2_{L^2(\mathbb{R})}\\
=&-12g\varepsilon\int_{\mathbb{R}}(\partial_x u_\varepsilon)^3\partial_x^6 u_\varepsilon dx-36g\varepsilon\int_{\mathbb{R}}u_\varepsilon\partial_x u_\varepsilon\partial_x^2 u_\varepsilon\partial_x^6 u_\varepsilon dx\\
&-6g\varepsilon\int_{\mathbb{R}}u_\varepsilon^2\partial_x^3 u_\varepsilon\partial_x^6 u_\varepsilon dx-6q\varepsilon\int_{\mathbb{R}}\partial_x v_\varepsilon\partial_x^2 u_\varepsilon\partial_x^6 u_\varepsilon dx\\
&-6q\varepsilon\int_{\mathbb{R}}\partial_x^2 v_\varepsilon\partial_x u_\varepsilon\partial_x^6 u_\varepsilon dx-2q\varepsilon\int_{\mathbb{R}}u_\varepsilon\partial_x^3 v_\varepsilon\partial_x^6 u_\varepsilon dx\\
&-2q\varepsilon\int_{\mathbb{R}}v_\varepsilon\partial_x^3 u_\varepsilon\partial_x^6 u_\varepsilon dx.
\end{aligned}\tag{A23}$$

Due to (24), (41), (42), (43), (56), (57), (A5) and the Young inequality,

$$\begin{aligned}
12|g\varepsilon|\int_{\mathbb{R}}|\partial_x u_\varepsilon|^3|\partial_x^6 u_\varepsilon|dx&=12\int_{\mathbb{R}}\left|\frac{g(\partial_x u_\varepsilon)^3}{\sqrt{D_1}}\right|\left|\varepsilon\sqrt{D_1}\partial_x^6 u_\varepsilon\right|dx\\
&\le\frac{6g^2}{D_1}\int_{\mathbb{R}}(\partial_x u_\varepsilon)^6dx+6D_1\varepsilon^2\left\|\partial_x^6 u_\varepsilon(t,\cdot)\right\|^2_{L^2(\mathbb{R})}\\
&\le\frac{6g^2}{D_1}\left\|\partial_x u_\varepsilon\right\|^4_{L^\infty((0,T)\times\mathbb{R})}\left\|\partial_x u_\varepsilon(t,\cdot)\right\|^2_{L^2(\mathbb{R})}+6D_1\varepsilon^2\left\|\partial_x^6 u_\varepsilon(t,\cdot)\right\|^2_{L^2(\mathbb{R})}\\
&\le\frac{C(T)}{D_1}+6D_1\varepsilon^2\left\|\partial_x^6 u_\varepsilon(t,\cdot)\right\|^2_{L^2(\mathbb{R})},\\
|36g\varepsilon|\int_{\mathbb{R}}|u_\varepsilon\partial_x u_\varepsilon\partial_x^2 u_\varepsilon||\partial_x^6 u_\varepsilon|dx&=36\int_{\mathbb{R}}\left|\frac{gu_\varepsilon\partial_x u_\varepsilon\partial_x^2 u_\varepsilon}{\sqrt{D_1}}\right|\left|\sqrt{D_1}\varepsilon\partial_x^6 u_\varepsilon\right|dx\\
&\le\frac{18g^2}{D_1}\int_{\mathbb{R}}u_\varepsilon^2(\partial_x u_\varepsilon)^2(\partial_x^2 u_\varepsilon)^2dx+18D_1\varepsilon^2\left\|\partial_x^6 u_\varepsilon(t,\cdot)\right\|^2_{L^2(\mathbb{R})}\\
&\le\frac{18g^2}{D_1}\left\|u_\varepsilon\right\|^2_{L^\infty((0,T)\times\mathbb{R})}\left\|\partial_x u_\varepsilon\right\|_{L^\infty((0,T)\times\mathbb{R})}\left\|\partial_x^2 u_\varepsilon(t,\cdot)\right\|^2_{L^2(\mathbb{R})}+18D_1\varepsilon^2\left\|\partial_x^6 u_\varepsilon(t,\cdot)\right\|^2_{L^2(\mathbb{R})}\\
&\le\frac{C(T)}{D_1}+18D_1\varepsilon^2\left\|\partial_x^6 u_\varepsilon(t,\cdot)\right\|^2_{L^2(\mathbb{R})},\\
|6g\varepsilon|\int_{\mathbb{R}}|u_\varepsilon^2\partial_x^3 u_\varepsilon||\partial_x^6 u_\varepsilon|dx&=6\int_{\mathbb{R}}\left|\frac{gu_\varepsilon^2\partial_x^3 u_\varepsilon}{\sqrt{D_1}}\right|\left|\sqrt{D_1}\varepsilon\partial_x^6 u_\varepsilon\right|dx\\
&\le\frac{3g^2}{D_1}\int_{\mathbb{R}}u_\varepsilon^4(\partial_x^3 u_\varepsilon)^2dx+3D_1\varepsilon^2\left\|\partial_x^6 u_\varepsilon(t,\cdot)\right\|^2_{L^2(\mathbb{R})}\\
&\le\frac{3g^2}{D_1}\left\|u_\varepsilon\right\|^4_{L^\infty((0,T)\times\mathbb{R})}\left\|\partial_x^3 u_\varepsilon(t,\cdot)\right\|^2_{L^2(\mathbb{R})}+3D_1\varepsilon^2\left\|\partial_x^6 u_\varepsilon(t,\cdot)\right\|^2_{L^2(\mathbb{R})}\\
&\le\frac{C(T)}{D_1}+3D_1\varepsilon^2\left\|\partial_x^6 u_\varepsilon(t,\cdot)\right\|^2_{L^2(\mathbb{R})},\\
|6q\varepsilon|\int_{\mathbb{R}}|\partial_x v_\varepsilon\partial_x^2 u_\varepsilon||\partial_x^6 u_\varepsilon|dx&=6\int_{\mathbb{R}}\left|\frac{q\partial_x v_\varepsilon\partial_x^2 u_\varepsilon}{\sqrt{D_1}}\right|\left|\sqrt{D_1}\varepsilon\partial_x^6 u_\varepsilon\right|dx\\
&\le\frac{3q^2}{D_1}\int_{\mathbb{R}}(\partial_x v_\varepsilon)^2(\partial_x^2 u_\varepsilon)^2dx+3D_1\varepsilon^2\left\|\partial_x^6 u_\varepsilon(t,\cdot)\right\|^2_{L^2(\mathbb{R})}\\
&\le\frac{3q^2}{D_1}\left\|\partial_x v_\varepsilon(t,\cdot)\right\|^2_{L^\infty(\mathbb{R})}\left\|\partial_x^2 u_\varepsilon(t,\cdot)\right\|^2_{L^2(\mathbb{R})}+3D_1\varepsilon^2\left\|\partial_x^6 u_\varepsilon(t,\cdot)\right\|^2_{L^2(\mathbb{R})}\\
&\le\frac{C(T)}{D_1}+3D_1\varepsilon^2\left\|\partial_x^6 u_\varepsilon(t,\cdot)\right\|^2_{L^2(\mathbb{R})},
\end{aligned}$$

$$\begin{aligned}
|6q\varepsilon| \int_{\mathbb{R}} |\partial_x^2 v_\varepsilon \partial_x u_\varepsilon| |\partial_x^6 u_\varepsilon| dx &= 6 \int_{\mathbb{R}} \left| \frac{q \partial_x^2 v_\varepsilon \partial_x u_\varepsilon}{\sqrt{D_1}} \right| \left| \sqrt{D_1} \varepsilon \partial_x^6 u_\varepsilon \right| dx \\
&\le \frac{3q^2}{D_1} \int_{\mathbb{R}} (\partial_x^2 v_\varepsilon)^2 (\partial_x u_\varepsilon)^2 dx + 3D_1 \varepsilon^2 \left\| \partial_x^6 u_\varepsilon(t,\cdot) \right\|^2_{L^2(\mathbb{R})} \\
&\le \frac{3q^2}{D_1} \left\| \partial_x^2 v_\varepsilon \right\|^2_{L^\infty((0,T)\times\mathbb{R})} \|\partial_x u_\varepsilon(t,\cdot)\|^2_{L^2(\mathbb{R})} + 3D_1 \varepsilon^2 \left\| \partial_x^6 u_\varepsilon(t,\cdot) \right\|^2_{L^2(\mathbb{R})} \\
&\le \frac{C(T)}{D_1} + 3D_1 \varepsilon^2 \left\| \partial_x^6 u_\varepsilon(t,\cdot) \right\|^2_{L^2(\mathbb{R})}, \\
|2q\varepsilon| \int_{\mathbb{R}} |u_\varepsilon \partial_x^3 v_\varepsilon| |\partial_x^6 u_\varepsilon| dx &= 2 \int_{\mathbb{R}} \left| \frac{q u_\varepsilon \partial_x^3 v_\varepsilon}{\sqrt{D_1}} \right| \left| \sqrt{D_1} \varepsilon \partial_x^6 u_\varepsilon \right| dx \\
&\le \frac{q^2}{D_1} \int_{\mathbb{R}} u_\varepsilon^2 (\partial_x^3 v_\varepsilon)^2 dx + D_1 \varepsilon^2 \left\| \partial_x^6 u_\varepsilon(t,\cdot) \right\|^2_{L^2(\mathbb{R})} \\
&\le \frac{q^2}{D_1} \|u_\varepsilon\|^2_{L^\infty((0,T)\times\mathbb{R})} \left\| \partial_x^3 v_\varepsilon(t,\cdot) \right\|^2_{L^2(\mathbb{R})} + D_1 \varepsilon^2 \left\| \partial_x^6 u_\varepsilon(t,\cdot) \right\|^2_{L^2(\mathbb{R})} \\
&\le \frac{C(T)}{D_1} + D_1 \varepsilon^2 \left\| \partial_x^6 u_\varepsilon(t,\cdot) \right\|^2_{L^2(\mathbb{R})}, \\
|2q\varepsilon| \int_{\mathbb{R}} |v_\varepsilon \partial_x^3 u_\varepsilon| |\partial_x^6 u_\varepsilon| dx &= \int_{\mathbb{R}} \left| \frac{q v_\varepsilon \partial_x^3 u_\varepsilon}{\sqrt{D_1}} \right| \left| \sqrt{D_1} \varepsilon \partial_x^6 u_\varepsilon \right| dx \\
&\le \frac{q^2}{D_1} \int_{\mathbb{R}} v_\varepsilon^2 (\partial_x^3 u_\varepsilon)^2 dx + D_1 \varepsilon^2 \left\| \partial_x^6 u_\varepsilon(t,\cdot) \right\|^2_{L^2(\mathbb{R})} \\
&\le \frac{q^2}{D_1} \|v_\varepsilon(t,\cdot)\|^2_{L^\infty(\mathbb{R})} \left\| \partial_x^3 u_\varepsilon(t,\cdot) \right\|^2_{L^2(\mathbb{R})} + D_1 \varepsilon^2 \left\| \partial_x^6 u_\varepsilon(t,\cdot) \right\|^2_{L^2(\mathbb{R})} \\
&\le \frac{C(T)}{D_1} + D_1 \varepsilon^2 \left\| \partial_x^6 u_\varepsilon(t,\cdot) \right\|^2_{L^2(\mathbb{R})},
\end{aligned}$$

where $D_1$ is a positive constant, which will be specified later. Consequently, by (A23),

$$\varepsilon \frac{d}{dt} \left\| \partial_x^4 u_\varepsilon(t,\cdot) \right\|^2_{L^2(\mathbb{R})} + (2 - 35D_1)\, \varepsilon^2 \left\| \partial_x^6 u_\varepsilon(t,\cdot) \right\|^2_{L^2(\mathbb{R})} \le \frac{C(T)}{D_1}.$$

Taking $D_1 = \frac{1}{35}$, we have that

$$\varepsilon \frac{d}{dt} \left\| \partial_x^4 u_\varepsilon(t,\cdot) \right\|^2_{L^2(\mathbb{R})} + \varepsilon^2 \left\| \partial_x^6 u_\varepsilon(t,\cdot) \right\|^2_{L^2(\mathbb{R})} \le C(T).$$

Equation (A3) and an integration on $(0,t)$ give

$$\varepsilon \left\| \partial_x^4 u_\varepsilon(t,\cdot) \right\|^2_{L^2(\mathbb{R})} + \varepsilon^2 \int_0^t \left\| \partial_x^6 u_\varepsilon(s,\cdot) \right\|^2_{L^2(\mathbb{R})} ds \le C_0 + C(T)t \le C(T),$$

that is (A17). □

**Lemma A4.** *Assume* (5)*. Fix $T > 0$. There exists a constant $C(T) > 0$, independent on $\varepsilon$, such that,*

$$\|\partial_t u_\varepsilon(t,\cdot)\|_{L^2(\mathbb{R})} \le C(T), \tag{A24}$$

*for every $0 \le t \le T$.*

**Proof.** Let $0 \le t \le T$. Multiplying the first equation of (16) by $2\partial_t u_\varepsilon$, an integration on $\mathbb{R}$ gives

$$
\begin{aligned}
2\left\|\partial_t u_\varepsilon(t,\cdot)\right\|^2_{L^2(\mathbb{R})} =& 2b\int_{\mathbb{R}} P_\varepsilon \partial_t u_\varepsilon dx - 2\varepsilon \int_{\mathbb{R}} \partial_x^4 u_\varepsilon \partial_t u_\varepsilon dx - 6g\int_{\mathbb{R}} u_\varepsilon^2 \partial_x u_\varepsilon \partial_t u_\varepsilon dx\\
&+ 2a\int_{\mathbb{R}} \partial_x^3 u_\varepsilon \partial_t u_\varepsilon dx - 2q\int_{\mathbb{R}} u_\varepsilon \partial_x v_\varepsilon \partial_t u_\varepsilon dx\\
&- 2q\int_{\mathbb{R}} v_\varepsilon \partial_x u_\varepsilon \partial_t u_\varepsilon dx.
\end{aligned}
\tag{A25}
$$

Since $0 < \varepsilon < 1$, thanks to (24), (41), (42), (43), (A5), (A17) and the Young inequality,

$$
\begin{aligned}
|2b|\int_{\mathbb{R}} |P_\varepsilon||\partial_t u_\varepsilon| dx =& 2\int_{\mathbb{R}} \left|\frac{bP_\varepsilon}{\sqrt{D_2}}\right| \left|\sqrt{D_2}\partial_t u_\varepsilon\right| dx\\
\le& \frac{b^2}{D_2}\left\|P_\varepsilon(t,\cdot)\right\|^2_{L^2(\mathbb{R})} + D_2\left\|\partial_t u_\varepsilon(t,\cdot)\right\|^2_{L^2(\mathbb{R})}\\
\le& \frac{C(T)}{D_2} + D_2\left\|\partial_t u_\varepsilon(t,\cdot)\right\|^2_{L^2(\mathbb{R})},\\
2\varepsilon\int_{\mathbb{R}} |\partial_x^4 u_\varepsilon||\partial_t u_\varepsilon| dx =& 2\int_{\mathbb{R}} \left|\frac{\varepsilon\partial_x^4 u_\varepsilon}{\sqrt{D_2}}\right| \left|\sqrt{D_2}\partial_t u_\varepsilon\right| dx\\
\le& \frac{\varepsilon^2}{D_2}\left\|\partial_x^4 u_\varepsilon(t,\cdot)\right\|^2_{L^2(\mathbb{R})} + D_2\left\|\partial_t u_\varepsilon(t,\cdot)\right\|^2_{L^2(\mathbb{R})}\\
\le& \frac{\varepsilon}{D_2}\left\|\partial_x^4 u_\varepsilon(t,\cdot)\right\|^2_{L^2(\mathbb{R})} + D_2\left\|\partial_t u_\varepsilon(t,\cdot)\right\|^2_{L^2(\mathbb{R})}\\
\le& \frac{C(T)}{D_2} + D_2\left\|\partial_t u_\varepsilon(t,\cdot)\right\|^2_{L^2(\mathbb{R})},\\
|6g|\int_{\mathbb{R}} |u_\varepsilon^2\partial_x u_\varepsilon||\partial_t u_\varepsilon| dx =& 6\int_{\mathbb{R}} \left|\frac{g u_\varepsilon^2\partial_x u_\varepsilon}{\sqrt{D_2}}\right| \left|\sqrt{D_2}\partial_t u_\varepsilon\right| dx\\
\le& \frac{3g^2}{D_2}\int_{\mathbb{R}} u_\varepsilon^4(\partial_x u_\varepsilon)^2 dx + 3D_2\left\|\partial_t u_\varepsilon(t,\cdot)\right\|^2_{L^2(\mathbb{R})}\\
\le& \frac{3g^2}{D_2}\left\|u_\varepsilon\right\|^4_{L^\infty((0,T)\times\mathbb{R})}\left\|\partial_x u_\varepsilon(t,\cdot)\right\|^2_{L^2(\mathbb{R})} + 3D_2\left\|\partial_t u_\varepsilon(t,\cdot)\right\|^2_{L^2(\mathbb{R})}\\
\le& \frac{C(T)}{D_2} + 3D_2\left\|\partial_t u_\varepsilon(t,\cdot)\right\|^2_{L^2(\mathbb{R})},\\
|2a|\int_{\mathbb{R}} |\partial_x^3 u_\varepsilon||\partial_t u_\varepsilon| dx =& 2\int_{\mathbb{R}} \left|\frac{a\partial_x^3 u_\varepsilon}{\sqrt{D_2}}\right| \left|\sqrt{D_2}\partial_t u_\varepsilon\right| dx\\
\le& \frac{a^2}{D_2}\left\|\partial_x^3 u_\varepsilon(t,\cdot)\right\|^2_{L^2(\mathbb{R})} + D_2\left\|\partial_t u_\varepsilon(t,\cdot)\right\|^2_{L^2(\mathbb{R})}\\
\le& \frac{C(T)}{D_2} + D_2\left\|\partial_t u_\varepsilon(t,\cdot)\right\|^2_{L^2(\mathbb{R})},\\
|2q|\int_{\mathbb{R}} |u_\varepsilon\partial_x v_\varepsilon||\partial_t u_\varepsilon| dx =& 2\int_{\mathbb{R}} \left|\frac{q u_\varepsilon\partial_x v_\varepsilon}{\sqrt{D_2}}\right| \left|\sqrt{D_2}\partial_t u_\varepsilon\right| dx\\
\le& \frac{q^2}{D_2}\int_{\mathbb{R}} u_\varepsilon^2(\partial_x v_\varepsilon)^2 dx + D_2\left\|\partial_t u_\varepsilon(t,\cdot)\right\|^2_{L^2(\mathbb{R})}\\
\le& \frac{q^2}{D_2}\left\|u_\varepsilon\right\|^2_{L^\infty((0,T)\times\mathbb{R})}\left\|\partial_x v_\varepsilon(t,\cdot)\right\|^2_{L^2(\mathbb{R})} + D_2\left\|\partial_t u_\varepsilon(t,\cdot)\right\|^2_{L^2(\mathbb{R})}\\
\le& \frac{C(T)}{D_2} + D_2\left\|\partial_t u_\varepsilon(t,\cdot)\right\|^2_{L^2(\mathbb{R})},
\end{aligned}
$$

$$
\begin{aligned}
|2q|\int_{\mathbb{R}}|v_\varepsilon\partial_x u_\varepsilon||\partial_t u_\varepsilon|dx &= 2\int_{\mathbb{R}}\left|\frac{qv_\varepsilon\partial_x u_\varepsilon}{\sqrt{D_2}}\right|\left|\sqrt{D_2}\partial_t u_\varepsilon\right|dx\\
&\le \frac{q^2}{D_2}\int_{\mathbb{R}}v_\varepsilon^2(\partial_x u_\varepsilon)^2dx + D_2\left\|\partial_t u_\varepsilon(t,\cdot)\right\|^2_{L^2(\mathbb{R})}\\
&\le \frac{q^2}{D_2}\left\|v_\varepsilon(t,\cdot)\right\|^2_{L^\infty(\mathbb{R})}\left\|\partial_x u_\varepsilon(t,\cdot)\right\|^2_{L^2(\mathbb{R})} + D_2\left\|\partial_t u_\varepsilon(t,\cdot)\right\|^2_{L^2(\mathbb{R})}\\
&\le \frac{C(T)}{D_2} + D_2\left\|\partial_t u_\varepsilon(t,\cdot)\right\|^2_{L^2(\mathbb{R})},
\end{aligned}
$$

where $D_2$ is a positive constant, which will be specified later. Therefore, by (A25),

$$
2\left(1-4D_2\right)\left\|\partial_t u_\varepsilon(t,\cdot)\right\|^2_{L^2(\mathbb{R})} \le \frac{C(T)}{D_2}.
$$

Choosing $D_2 = \frac{1}{8}$, we have that

$$
\left\|\partial_t u_\varepsilon(t,\cdot)\right\|^2_{L^2(\mathbb{R})} \le C(T),
$$

which gives (A24). □

Arguing as in ([15], Lemma 2.12), we have the following result.

**Lemma A5.** *Assume* (5). *Let $T > 0$. There exists a constant $C(T) > 0$, independent on $\varepsilon$, such that*

$$
\begin{aligned}
\left\|\partial^2_{tx}v_\varepsilon(t,\cdot)\right\|_{L^\infty(\mathbb{R})}, \left\|\partial^2_{tx}v_\varepsilon(t,\cdot)\right\|_{L^2(\mathbb{R})} &\le C(T),\\
\left\|\partial_t v_\varepsilon(t,\cdot)\right\|_{L^\infty(\mathbb{R})}, \left\|\partial_t v_\varepsilon(t,\cdot)\right\|_{L^2(\mathbb{R})} &\le C(T),
\end{aligned}
\tag{A26}
$$

*for every $0 \le t \le T$.*

Using the Sobolev Immersion Theorem, we begin by proving the following result.

**Lemma A6.** *Fix $T > 0$. There exist a subsequence $\{(u_{\varepsilon_k}, v_{\varepsilon_k}, P_{\varepsilon_k})\}_{k\in\mathbb{N}}$ of $\{(u_\varepsilon, v_\varepsilon, P_\varepsilon)\}_{\varepsilon>0}$ and an a limit triplet $(u, v, P)$ which satisfies* (11) *such that*

$$
\begin{aligned}
&u_{\varepsilon_k} \to u \text{ a.e. and in } L^p_{loc}((0,T)\times\mathbb{R}), 1 \le p < \infty,\\
&u_{\varepsilon_k} \rightharpoonup u \text{ in } H^1((0,T)\times\mathbb{R}),\\
&v_{\varepsilon_k} \to v \text{ a.e. and in } L^p_{loc}((0,T)\times\mathbb{R}), 1 \le p < \infty,\\
&v_{\varepsilon_k} \rightharpoonup v \text{ in } H^1((0,T)\times\mathbb{R}),\\
&P_{\varepsilon_k} \rightharpoonup P \text{ in } L^2((0,T)\times\mathbb{R}).
\end{aligned}
\tag{A27}
$$

*Moreover, $(u, v, P)$ is solution of* (1) *satisfying* (12).

**Proof.** Let $0 \le t \le T$. We begin by observing that, thanks to Lemmas 3, 7, 9, A1 and A4,

$$
\{u_\varepsilon\}_{\varepsilon>0} \text{ is uniformly bounded in } H^1((0,T)\times\mathbb{R}). \tag{A28}
$$

Lemmas 3 and A5 say that

$$
\{v_\varepsilon\}_{\varepsilon>0} \text{ is uniformly bounded in } H^1((0,T)\times\mathbb{R}). \tag{A29}
$$

Instead, by Lemma 7, we have that

$$\{P_\varepsilon\}_{\varepsilon>0} \text{ is uniformly bounded in } L^2((0,T)\times\mathbb{R}). \tag{A30}$$

Equation (A28), (A29) and (A30) give (A27).

Observe that, thanks to Lemmas 3, 7, 9, A1 and the second equation of (16), we have that

$$P \in L^\infty(0,T;H^4(\mathbb{R})).$$

Lemmas 3, 7, 9, A1 say that

$$u \in L^\infty(0,T;H^3(\mathbb{R})).$$

Instead, thanks to Lemmas 3, 8, 10, A2 and (A26), we get

$$v \in L^\infty(0,T;H^5(\mathbb{R})) \cap W^{1,\infty}((0,T)\times\mathbb{R}).$$

Moreover, Lemmas A5 says also that

$$\partial_{tx}^2 v \in L^2(\mathbb{R}) \cap L^\infty(\mathbb{R}),$$

for every $0 \le t \le T$. Therefore, (11) holds and $(u,\, v,\, P)$ is solution of (1).

Finally, (12) follows from (19) and (A27). □

Now, we prove Theorem A1.

**Proof of Theorem A1.** Lemma A6 gives the existence of a solution of (1) such that (12) and (A27) hold. Arguing as in Theorem 1, we have (13). □

## References

1. Coclite, G.M.; di Ruvo, L. Convergence of the Ostrovsky Equation to the Ostrovsky-Hunter One. *J. Differ. Equ.* **2014**, *256*, 3245–3277. [CrossRef]
2. Coclite, G.M.; di Ruvo, L. Oleinik type estimate for the Ostrovsky-Hunter equation. *J. Math. Anal. Appl.* **2015**, *423*, 162–190. [CrossRef]
3. Coclite, G.M.; di Ruvo, L. Wellposedness of bounded solutions of the non-homogeneous initial boundary value problem for the Ostrovsky-Hunter equation. *J. Hyperbolic Differ. Equ.* **2015**, *12*, 221–248. [CrossRef]
4. Coclite, G.M.; di Ruvo, L. Wellposedness results for the short pulse equation. *Z. Angew. Math. Phys.* **2015**, *66*, 1529–1557. [CrossRef]
5. Coclite, G.M.; di Ruvo, L. Well-posedness and dispersive/diffusive limit of a generalized Ostrovsky-Hunter equation. *Milan J. Math.* **2018**, *86*, 31–51. [CrossRef]
6. Coclite, G.M.; di Ruvo, L. Classical solutions for an Ostrovsky type equation. Submitted.
7. Bespalov, V.G.; Kozlov, S.A.; Shpolyanskiy, Y.A. Method for analyzing the propagation dynamics of femtosecond pulses with a continuum spectrum in transparent optical media. *J. Opt. Technol.* **2000**, *67*, 5–11. [CrossRef]
8. Bespalov, V.G.; Kozlov, S.A.; Shpolyanskiy, Y.A.; Walmsley, I.A. Simplified field wave equations for the nonlinear propagation of extremely short light pulses. *Phys. Rev. A* **2002**, *66*, 013811. [CrossRef]
9. Bespalov, V.G.; Kozlov, S.A.; Sutyagin, A.N.; Shpolyansky, Y.A. Spectral super-broadening of high-power femtosecond laser pulses and their time compression down to one period of the light field. *J. Opt. Technol.* **1998**, *65*, 823–825.
10. Gagarskiĭ, S.V.; Prikhod'ko, K.V. Measuring the parameters of femtosecond pulses in a wide spectral range on the basis of the multiphoton-absorption effect in a natural diamond crystal. *J. Opt. Technol.* **2008**, *75*, 139–143. [CrossRef]
11. Konev, L.S.; Shpolyanskiĭ, Y.A. Calculating the field and spectrum of the reverse wave induced when a femtosecond pulse with a superwide spectrum propagates in an optical waveguide. *J. Opt. Technol.* **2014**, *81*, 6–11. [CrossRef]
12. Kozlov, S.A.; Sazonov, S.V. Nonlinear propagation of optical pulses of a few oscillations duration in dielectric media. *J. Exp. Theor. Phys.* **1997**, *84*, 221–228. [CrossRef]

13. Melnik, M.V.; Tcypkin, A.N.; Kozlov, S.A. Temporal coherence of optical supercontinuum. *Rom. J. Phys.* **2018**, *63*, 203.
14. Shpolyanskiy, Y.A.; Belov, D.I.; Bakhtin, M.A.; Kozlov, S.A. Analytic study of continuum spectrum pulse dynamics in optical waveguides. *Appl. Phys. B* **2003**, *77*, 349–355. [CrossRef]
15. Coclite, G.M.; di Ruvo, L. A non-local elliptic-hyperbolic system related to the short pulse equation. *Nonlinear Anal.* **2020**, *190*, 111606. [CrossRef]
16. Coclite, G.M.; di Ruvo, L. Convergence of the solutions on the generalized Korteweg-de Vries equation. *Math. Model. Anal.* **2016**, *21*, 239–259. [CrossRef]
17. Colliander, J.; Keel, M.; Staffilani, G.; Takaoka, H.; Tao, T. Sharp global wellposednees for KDV and modified KDV on $\mathbb{R}$ and $\mathbb{T}$. *J. Am. Math. Soc.* **2003**, *16*, 705–749. [CrossRef]
18. Kenig, C.E.; Ponce, G.; Vega, L. Wellposedness and scattering results for the generalized Korteweg-de Vries Equation via the contraction principle. *Commun. Pure Appl. Math.* **1993**, *46*, 527–620. [CrossRef]
19. Schonbek, M.E. Convergence of solutions to nonlinear dispersive equations. *Commun. Part. Differ. Equ.* **1982**, *7*, 959–1000.
20. Tao, T. Nonlinear dispersive equations. In *CBMS Regional Conference Series in Mathematics*; Local and Global Analysis; Conference Board of the Mathematical Sciences: Washington, DC, USA, 2006; Volume 106.
21. Belashenkov, N.R.; Drozdov, A.A.; Kozlov, S.A.; Shpolyanskiy, Y.A.; Tsypkin, A.N. Phase modulation of femtosecond light pulses whose spectra are superbroadened in dielectrics with normal group dispersion. *J. Opt. Technol.* **2008**, *75*, 611–614. [CrossRef]
22. Leblond, H.; Mihalache, D. Few-optical-cycle solitons: Modified Korteweg-de Vries sine-Gordon equation versus other non-slowly-varying-envelope-approximation models. *Phys. Rev. A* **2009**, *79*, 063835. [CrossRef]
23. Leblond, H.; Mihalache, D. Models of few optical cycle solitons beyond the slowly varying envelope approximation. *Phys. Rep.* **2013**, *523*, 61–126. [CrossRef]
24. Leblond, H.; Sanchez, F. Models for optical solitons in the two-cycle regime. *Phys. Rev. A* **2003**, *67*, 013804. [CrossRef]
25. Schäfer, T.; Wayne, C.E. Propagation of ultra-short optical pulses in cubic nonlinear media. *Physica D* **2004**, *196*, 90–105. [CrossRef]
26. Amiranashvili, S.; Vladimirov, A.G.; Bandelow, U. A model equation for ultrashort optical pulses. *Eur. Phys. J. D* **2010**, *58*, 219. [CrossRef]
27. Amiranashvili, S.; Vladimirov, A.G.; Bandelow, U. Solitary-wave solutions for few-cycle optical pulses. *Phys. Rev. A* **2008**, *77*, 063821. [CrossRef]
28. Coclite, G.M.; di Ruvo, L. Discontinuous solutions for the generalized short pulse equation. *Evol. Equ. Control Theory* **2019**, *8*, 737–753. [CrossRef]
29. Beals, R.; Rabelo, M.; Tenenblat, K. Bäcklund transformations and inverse scattering solutions for some pseudospherical surface equations. *Stud. Appl. Math.* **1989**, *81*, 125–151. [CrossRef]
30. Coclite, G.M.; di Ruvo, L. On the well-posedness of the exp-Rabelo equation. *Ann. Mater. Pure Appl.* **2016**, *195*, 923–933. [CrossRef]
31. Rabelo, M. On equations which describe pseudospherical surfaces. *Stud. Appl. Math.* **1989**, *81*, 221–248. [CrossRef]
32. Sakovich, A.; Sakovich, S. On the transformations of the Rabelo equations. *SIGMA 3* **2007**, 8. [CrossRef]
33. Tsitsas, N.L.; Horikis, T.P.; Shen, Y.; Kevrekidis, P.G.; Whitaker, N.; Frantzeskakis, D.J. Short pulse equations and localized structures in frequency band gaps of nonlinear metamaterials. *Phys. Lett. A* **2010**, *374*, 1384–1388. [CrossRef]
34. Nikitenkova, S.P.; Stepanyants, Y.A.; Chikhladze, L.M. Solutions of the modified Ostrovskii equation with cubic non-linearity. *J. Appl. Math. Mech.* **2000**, *64*, 267–274. [CrossRef]
35. Lattanzio, C.; Marcati, P. Global well-posedness and relaxation limits of a model for radiating gas. *J. Differ. Equ.* **2013**, *190*, 439–465. [CrossRef]
36. Serre, D. $L^1$-stability of constants in a model for radiating gases. *Commun. Math. Sci.* **2003**, *1*, 197–205. [CrossRef]
37. Coclite, G.M.; di Ruvo, L. Discontinuous solutions for the short-pulse master mode-locking equation. *AIMS Math.* **2019**, *4*, 437–462. [CrossRef]
38. Farnum, E.D.; Kutz, J.N. Master mode-locking theory for few-femtosecond pulses. *J. Opt. Soc. Am. B* **2010**, *35*, 3033–3035. [CrossRef]

39. Farnum, E.D.; Kutz, J.N. Dynamics of a low-dimensional model for short pulse mode locking. *Photonics* **2015**, *2*, 865–882. [CrossRef]
40. Farnum, E.D.; Kutz, J.N. Short-pulse perturbation theory. *J. Opt. Soc. Am. B* **2013**, *30*, 2191–2198. [CrossRef]
41. Pelinovsky, D.; Schneider, G. Rigorous justification of the short-pulse equation. *Nonlinear Differ. Equ. Appl.* **2013**, *20*, 1277–1294. [CrossRef]
42. Davidson, M. Continuity properties of the solution map for the generalized reduced Ostrovsky equation. *J. Differ. Equ.* **2012**, *252*, 3797–3815. [CrossRef]
43. Pelinovsky, D.; Sakovich, A. Global well-posedness of the short-pulse and sine-Gordon equations in energy space. *Commun. Part. Differ. Equ.* **2010**, *352*, 613–629. [CrossRef]
44. Stefanov, A.; Shen, Y.; Kevrekidis, P.G. Well-posedness and small data scattering for the generalized Ostrovsky equation. *J. Differ. Equ.* **2010**, *249*, 2600–2617. [CrossRef]
45. Coclite, G.M.; di Ruvo, L.; Karlsen, K.H. Some wellposedness results for the Ostrovsky-Hunter equation. In *Hyperbolic Conservation Laws and Related Analysis with Applications*; Springer Proceedings in Mathematics & Statistics; Springer: Heidelberg, Germany, 2014; Volume 49, pp. 143–159.
46. Di Ruvo, L. Discontinuous Solutions for the Ostrovsky–Hunter Equation and Two Phase Flows. Ph.D. Thesis, University of Bari, Bari, Italy, 2013. Avalaible online: www.dm.uniba.it/home/dottorato/dottorato/tesi/ (accessed on 1 June 2013).
47. Coclite, G.M.; di Ruvo, L. Wellposedness of bounded solutions of the non-homogeneous initial boundary for the short pulse equation. *Boll. Unione Mater. Ital.* **2015**, *8*, 31–44. [CrossRef]
48. Coclite, G.M.; di Ruvo, L. A note on the non-homogeneous initial boundary problem for an Ostrovsky-Hunter type equation. *Discr. Contin. Dyn. Syst. Ser. S*. To Appear.
49. Coclite, G.M.; di Ruvo, L.; Karlsen, K.H. The initial-boundary-value problem for an Ostrovsky-Hunter type equation. In *Non-Linear Partial Differential Equations, Mathematical Physics, and Stochastic Analysis*; EMS Series Congress Report; European Mathematical Society Zürich, Zürich, Switzerland; 2018; pp. 97–109.
50. Ridder, J.; Ruf, A.M. A convergent finite difference scheme for the Ostrovsky-Hunter equation with Dirichlet boundary conditions. *BIT Numer. Math.* **2019**, *59*, 775–796. [CrossRef]
51. Coclite, G.M.; di Ruvo, L. A non-local regularization of the short pulse equation. *Minimax Theory Appl.*
52. Coclite, G.M.; Ridder, J.; Risebro, H. A convergent finite difference scheme for the Ostrovsky-Hunter equation on a bounded domain. *BIT Numer. Math.* **2017**, *57*, 93–122. [CrossRef]
53. Costanzino, N.; Manukian, V.; Jones, C.K.R.T. Solitary waves of the regularized short pulse and Ostrovsky equations. *SIAM J. Math. Anal.* **2009**, *41*, 2088–2106. [CrossRef]
54. Liu, Y.; Pelinovsky, D.; Sakovich, A. Wave breaking in the short-pulse equation. *Dyn. PDE* **2009**, *6*, 291–310. [CrossRef]
55. Coclite, G.M.; di Ruvo, L. Dispersive and Diffusive limits for Ostrovsky-Hunter type equations. *Nonlinear Differ. Equ. Appl.* **2015**, *22*, 1733–1763. [CrossRef]
56. LeFloch, P.G.; Natalini, R. Conservation laws with vanishing nonlinear diffusion and dispersion. *Nonlinear Anal.* **1992**, *36*, 212–230.
57. Coclite, G.M.; di Ruvo, L. Convergence of the regularized short pulse equation to the short pulse one. *Math. Nachr.* **2018**, *291*, 774–792. [CrossRef]
58. Coclite, G.M.; Garavello, M. A Time Dependent Optimal Harvesting Problem with Measure Valued Solutions. *SIAM J. Control Optim.* **2017**, *55*, 913–935. [CrossRef]
59. Coclite, G.M.; Garavello, M.; Spinolo, L.V. Optimal strategies for a time-dependent harvesting problem. *Discr. Contin. Dyn. Syst. Ser. S* **2016**, *11*, 865–900. [CrossRef]
60. Simon, J. Compact sets in the space $L_p(0,T;B)$. *Ann. Mater. Pura Appl.* **1987**, *4*, 65–94.
61. Coclite, G.M.; di Ruvo, L. Wellposedness of the Ostrovsky-Hunter Equation under the combined effects of dissipation and short wave dispersion. *J. Evol. Equ.* **2016**, *16*, 365–389. [CrossRef]
62. Coclite, G.M.; Holden, H.; Karlsen, K.H. Wellposedness for a parabolic-elliptic system. *Discr. Contin. Dyn. Syst.* **2005**, *13*, 659–682.

Article

# Strong Solutions of the Incompressible Navier–Stokes–Voigt Model

**Evgenii S. Baranovskii**

Department of Applied Mathematics, Informatics and Mechanics, Voronezh State University, 394018 Voronezh, Russia; esbaranovskii@gmail.com

Received: 30 December 2019; Accepted: 28 January 2020; Published: 3 February 2020

**Abstract:** This paper deals with an initial-boundary value problem for the Navier–Stokes–Voigt equations describing unsteady flows of an incompressible non-Newtonian fluid. We give the strong formulation of this problem as a nonlinear evolutionary equation in Sobolev spaces. Using the Faedo–Galerkin method with a special basis of eigenfunctions of the Stokes operator, we construct a global-in-time strong solution, which is unique in both two-dimensional and three-dimensional domains. We also study the long-time asymptotic behavior of the velocity field under the assumption that the external forces field is conservative.

**Keywords:** Navier–Stokes–Voigt equations; viscoelastic models; non-Newtonian fluid; strong solutions; existence and uniqueness theorem; Faedo–Galerkin approximations; Stokes operator; long-time behavior

---

## 1. Introduction

In this work, we study an initial-boundary value problem for the Navier–Stokes–Voigt (NSV) equations that model the unsteady flow of an incompressible viscoelastic fluid:

$$\begin{cases} \dfrac{\partial \boldsymbol{u}}{\partial t} + (\boldsymbol{u}\cdot\nabla)\boldsymbol{u} - \nu\Delta\boldsymbol{u} - \alpha^2\dfrac{\partial\Delta\boldsymbol{u}}{\partial t} + \nabla p = \boldsymbol{f} \quad \text{in } \Omega\times(0,+\infty), \\ \nabla\cdot\boldsymbol{u} = 0 \quad \text{in } \Omega\times(0,+\infty), \\ \boldsymbol{u} = \boldsymbol{0} \quad \text{on } \partial\Omega\times(0,+\infty), \\ \boldsymbol{u}(\cdot,0) = \boldsymbol{u}_0 \quad \text{in } \Omega, \end{cases} \tag{1}$$

where $\Omega$ denotes the bounded domain of flow in $\mathbb{R}^n$, $n = 2,3$, with boundary $\partial\Omega$; the vector function $\boldsymbol{u}$ represents the velocity field; $p$ denotes the pressure; $\nu > 0$ is the viscosity coefficient; $\alpha$ is a length scale parameter such that $\alpha^2/\nu$ is the relaxation time of the viscoelastic fluid; $\boldsymbol{f}$ is the external forces field; and $\boldsymbol{u}_0$ is the initial velocity.

Note that when $\alpha = 0$ the NSV system becomes the incompressible Navier–Stokes equations that describe Newtonian fluid flows. If $\alpha = 0$ and $\nu = 0$, then we arrive at the incompressible Euler equations governing inviscid flows.

In the literature, the NSV equations are often called the Kelvin–Voigt equations or Oskolkov's equations. The NSV model and related models of viscoelastic fluid flows have been studied extensively by different mathematicians over the past several decades starting from the pioneering papers by Oskolkov [1,2]. It should be mentioned at this point that Oskolkov later admitted [3] that these works contain some errors and not all obtained results hold. In this regard, Ladyzhenskaya remarked in her note [4] that the method of introduction of auxiliary viscosity used in [1,2] is incorrect under the no-slip boundary condition and explained the reasons for this. However, it is certain that the series of Oskolkov's works played a major role in the study of the NSV equations and stimulated further research in this direction.

Let us shortly review available literature on mathematical analysis of NSV-type models. Sviridyuk [5] established the solvability of the weakly compressible NSV equations. In [6], the local-in-time unique solvability of problem (1) is proved. Korpusov and Sveshnikov [7] investigated the blowup of solutions to the NSV equations with a cubic source. Various slip problems are studied in the papers [8–10]. Kaya and Celebi [11] proved the existence and uniqueness of weak solutions of the so-called g-Kelvin–Voigt equations that describe viscoelastic fluid flows in thin domains. The solvability of the inhomogeneous Dirichlet problem for the equations governing a polymer fluid flow is proved in [12]. Berselli and Spirito [13] showed that weak solutions to the Navier–Stokes equations obtained as limits $\alpha \to 0^+$ of solutions to the NSV model are "suitable weak solutions" [14] and satisfy the local energy inequality. Fedorov and Ivanova [15] dealt with an inverse problem for the NSV equations. An algorithm for finding of numerical solution of an optimal control problem for the two-dimensional Kelvin–Voigt fluid flow was proposed by Plekhanova et al. [16]. Antontsev and Khompysh [17] established the existence and uniqueness of the global and local weak solutions to the NSV equations with p-Laplacian and a damping term. Artemov and Baranovskii [18] proved the existence of weak solutions to the coupled system of nonlinear equations describing the heat transfer in steady-state flows of a polymeric fluid. Mohan [19] investigated the global solvability, the asymptotic behavior, and some control problems for the NSV model with "fading memory" and "memory of length $\tau$".

Most of the papers mentioned above deal with the study of weak (generalized) solutions to the NSV equations in the framework of the Hilbert space techniques. Therefore, it is a relevant question to prove the existence and uniqueness of strong solutions of system (1) in a Banach space under natural conditions on the data. Another important objective is to develop convenient algorithms for finding strong solutions or their approximations. Motivated by this, in the present work, we propose the strong formulation of problem (1) as a nonlinear evolutionary equation in suitable Banach spaces with the initial condition $\boldsymbol{u}(0) = \boldsymbol{u}_0$. Using the Faedo–Galerkin procedure with a special basis of eigenfunctions of the Stokes operator and deriving various a priori estimates of approximate solutions in Sobolev's spaces $\boldsymbol{H}^1(\Omega)$ and $\boldsymbol{H}^2(\Omega)$, we construct a global-in-time strong solution of (1), which is unique in both two-dimensional and three-dimensional domains. We also derive the energy equality that holds for strong solutions. Moreover, it is shown that, if the external forces field $\boldsymbol{f}$ is conservative, then the $H^1$-norm of the velocity field $\boldsymbol{u}$ decays exponentially as $t \to +\infty$.

## 2. Preliminaries

To suggest the concept of a strong solution to problem (1), we introduce some notations, function spaces, and auxiliary results.

For vectors $\boldsymbol{x}, \boldsymbol{y} \in \mathbb{R}^n$ and matrices $\mathbf{X}, \mathbf{Y} \in \mathbb{R}^{n\times n}$ by $\boldsymbol{x} \cdot \boldsymbol{y}$ and $\mathbf{X} : \mathbf{Y}$, we denote the scalar products, respectively:

$$\boldsymbol{x} \cdot \boldsymbol{y} \stackrel{\text{def}}{=} \sum_{i=1}^{n} x_i y_i, \qquad \mathbf{X} : \mathbf{Y} \stackrel{\text{def}}{=} \sum_{i,j=1}^{n} \mathrm{X}_{ij} \mathrm{Y}_{ij}.$$

Let $\Omega \subset \mathbb{R}^n$ be a bounded domain with sufficiently smooth boundary $\partial\Omega$. By $\mathcal{D}(\Omega)$ denote the set of $C^\infty$ functions with support contained in $\Omega$. We use the standard notation for the Lebesgue spaces $L^s(\Omega)$, $s \geq 1$, as well as the Sobolev spaces $H^k(\Omega) \stackrel{\text{def}}{=} W^{k,2}(\Omega)$, $k \in \mathbb{N}$. When it comes to classes of $\mathbb{R}^n$-valued functions, we employ boldface letters, for instance,

$$\boldsymbol{D}(\Omega) \stackrel{\text{def}}{=} \mathcal{D}(\Omega)^n, \qquad \boldsymbol{L}^s(\Omega) \stackrel{\text{def}}{=} L^s(\Omega)^n, \qquad \boldsymbol{H}^k(\Omega) \stackrel{\text{def}}{=} H^k(\Omega)^n.$$

It is well known that the space Sobolev $H^1(\Omega)$ is compactly embedded in $L^4(\Omega)$.

Let us introduce the following spaces:

$$\begin{aligned}
\mathcal{V}(\Omega) &\stackrel{\text{def}}{=} \{\boldsymbol{v} \in \boldsymbol{D}(\Omega)\colon\ \nabla \cdot \boldsymbol{v} = 0\},\\
\boldsymbol{V}^0(\Omega) &\stackrel{\text{def}}{=} \text{the closure of the set } \mathcal{V}(\Omega) \text{ in the space } \boldsymbol{L}^2(\Omega),\\
\boldsymbol{V}^1(\Omega) &\stackrel{\text{def}}{=} \text{the closure of the set } \mathcal{V}(\Omega) \text{ in the space } \boldsymbol{H}^1(\Omega),\\
\boldsymbol{V}^2(\Omega) &\stackrel{\text{def}}{=} \boldsymbol{H}^2(\Omega) \cap \boldsymbol{V}^1(\Omega).
\end{aligned} \tag{2}$$

It is obvious that $\boldsymbol{V}^0(\Omega)$, $\boldsymbol{V}^1(\Omega)$, and $\boldsymbol{V}^2(\Omega)$ are Hilbert spaces with the scalar products induced by $\boldsymbol{L}^2(\Omega)$, $\boldsymbol{H}^1(\Omega)$, and $\boldsymbol{H}^2(\Omega)$, respectively. However, when studying problem (1), in the spaces $\boldsymbol{V}^1(\Omega)$ and $\boldsymbol{V}^2(\Omega)$, it is more convenient to use the scalar products and the norms defined as follows:

$$(\boldsymbol{v},\boldsymbol{w})_{\boldsymbol{V}^1(\Omega)} \stackrel{\text{def}}{=} (\boldsymbol{v},\boldsymbol{w})_{\boldsymbol{L}^2(\Omega)} + \alpha^2(\nabla\boldsymbol{v},\nabla\boldsymbol{w})_{\boldsymbol{L}^2(\Omega)}, \qquad \|\boldsymbol{v}\|_{\boldsymbol{V}^1(\Omega)} \stackrel{\text{def}}{=} (\boldsymbol{v},\boldsymbol{v})^{1/2}_{\boldsymbol{V}^1(\Omega)}, \tag{3}$$

$$(\boldsymbol{v},\boldsymbol{w})_{\boldsymbol{V}^2(\Omega)} \stackrel{\text{def}}{=} (\mathbb{P}\Delta\boldsymbol{v},\mathbb{P}\Delta\boldsymbol{w})_{\boldsymbol{L}^2(\Omega)}, \qquad \|\boldsymbol{v}\|_{\boldsymbol{V}^2(\Omega)} \stackrel{\text{def}}{=} (\boldsymbol{v},\boldsymbol{v})^{1/2}_{\boldsymbol{V}^2(\Omega)}. \tag{4}$$

Here, $\mathbb{P}\colon \boldsymbol{L}^2(\Omega) \to \boldsymbol{V}^0(\Omega)$ is the Leray projection, which corresponds the well-known Leray (or Hodge–Helmholtz) decomposition for the vector fields in $\boldsymbol{L}^2(\Omega)$ into a divergence-free part and a gradient part (see, e.g., [20], Chapter IV):

$$\boldsymbol{L}^2(\Omega) = \boldsymbol{V}^0(\Omega) \oplus \boldsymbol{G}(\Omega),$$

where the symbol $\oplus$ denotes the orthogonal sum and the subspace $\boldsymbol{G}(\Omega)$ is defined as follows

$$\boldsymbol{G}(\Omega) \stackrel{\text{def}}{=} \{\nabla h:\ h \in H^1(\Omega)\}.$$

Note that the norm $\|\cdot\|_{\boldsymbol{V}^i(\Omega)}$ is equivalent to the norm $\|\cdot\|_{\boldsymbol{H}^i(\Omega)}$, $i = 1,2$.

We introduce the equivalence relation on the space $H^1(\Omega)$ by stating that $\varphi \sim \psi$ if $\varphi - \psi = \text{const}$. As usual, $H^1(\Omega)/\mathbb{R}$ denotes the quotient of $H^1(\Omega)$ by $\mathbb{R}$.

For a function $\xi \in H^1(\Omega)$, we set

$$\overline{\xi} \stackrel{\text{def}}{=} \{\omega \in H^1(\Omega):\ \omega \sim \xi\} \in H^1(\Omega)/\mathbb{R}.$$

Let us define the gradient and the norm of $\overline{\xi}$ as follows

$$\nabla\overline{\xi} \stackrel{\text{def}}{=} \nabla\xi, \qquad \|\overline{\xi}\|_{H^1(\Omega)/\mathbb{R}} \stackrel{\text{def}}{=} \|\nabla\overline{\xi}\|_{L^2(\Omega)}.$$

Using Proposition 1.2 from ([21], Chapter I, § 1), it is easy to verify that the norm $\|\cdot\|_{H^1(\Omega)/\mathbb{R}}$ is well defined.

The following lemmas are needed for the sequel.

**Lemma 1.** *Suppose $\boldsymbol{E}$ is a Banach space and $T$ is a positive number. A set $\boldsymbol{K}$ of the space $\boldsymbol{C}([0,T];\boldsymbol{E})$ is relatively compact if and only if:*

- *for any number $t \in (0,T)$, the set $\boldsymbol{K}(t) \stackrel{\text{def}}{=} \{\boldsymbol{w}(t):\ \boldsymbol{w} \in \boldsymbol{K}\}$ is relatively compact in $\boldsymbol{E}$;*
- *for any number $\varepsilon > 0$, there exists a number $\eta$ >0 such that the inequality*

$$\|\boldsymbol{w}(t_1) - \boldsymbol{w}(t_2)\|_{\boldsymbol{E}} \le \varepsilon$$

*holds for any function $\boldsymbol{w} \in \boldsymbol{K}$ and any numbers $t_1, t_2 \in [0,T]$ such that $|t_1 - t_2| \le \eta$.*

The proof of this lemma is given in [22].

**Lemma 2.** *The embedding $C^1([0,T];V^2(\Omega)) \hookrightarrow C([0,T];V^1(\Omega))$ is completely continuous.*

**Proof.** Let $S$ be a bounded set of $C^1([0,T];V^2(\Omega))$. Then

$$\max_{(w,t)\in S\times[0,T]} \|w(t)\|_{V^2(\Omega)} + \max_{(w,t)\in S\times[0,T]} \|w'(t)\|_{V^2(\Omega)} \leq r \tag{5}$$

with some constant $r$. Clearly, this implies that the set

$$S(t) \stackrel{\text{def}}{=} \{w(t) : w \in S\}$$

is bounded in $V^2(\Omega)$ for any $t \in [0,T]$.

From the Rellich–Kondrachov theorem (see, e.g., [23], Chapter 1, Theorem 1.12.1), it follows that the space $V^2(\Omega)$ is compactly embedded into $V^1(\Omega)$. Therefore, the set $S(t)$ is relatively compact in the space $V^1(\Omega)$.

By $\mathbf{I}$ denote the embedding operator from $V^2(\Omega)$ into $V^1(\Omega)$. Taking into account inequality (5), we get the estimate

$$\begin{aligned}\|w(t_1) - w(t_2)\|_{V^1(\Omega)} &\leq \|\mathbf{I}\|_{\mathcal{L}(V^2(\Omega),V^1(\Omega))}\|w(t_1) - w(t_2)\|_{V^2(\Omega)} \\ &\leq \|\mathbf{I}\|_{\mathcal{L}(V^2(\Omega),V^1(\Omega))} \max_{\tau\in[t_1,t_2]} \|w'(\tau)\|_{V^2(\Omega)}|t_1 - t_2| \\ &\leq r\|\mathbf{I}\|_{\mathcal{L}(V^2(\Omega),V^1(\Omega))}|t_1 - t_2| \leq \varepsilon\end{aligned}$$

for any function $w \in S$ and for any numbers $t_1, t_2 \in [0,T]$ such that

$$|t_1 - t_2| \leq \frac{\varepsilon}{r\|\mathbf{I}\|_{\mathcal{L}(V^2(\Omega),V^1(\Omega))}},$$

where $\|\mathbf{I}\|_{\mathcal{L}(V^2(\Omega),V^1(\Omega))}$ is the operator norm of $\mathbf{I}$.

Applying Lemma 1 with $E = V^1(\Omega)$, we conclude that the set $S$ is relatively compact in the space $C([0,T];V^1(\Omega))$. Lemma 2 is proved. □

**Lemma 3.** *Let*

$$\mathfrak{L}_d \stackrel{\text{def}}{=} \{x \in \mathbb{R}^n : |x_n| < d/2\}, \qquad d_\Omega \stackrel{\text{def}}{=} \inf\{d > 0 : \Omega \subset \mathfrak{L}_d\}.$$

*Then, we have*

$$\frac{4}{d_\Omega^2 + 4\alpha^2}\left(\|v\|^2_{L^2(\Omega)} + \alpha^2\|\nabla v\|^2_{L^2(\Omega)}\right) \leq \|\nabla v\|^2_{L^2(\Omega)}, \tag{6}$$

*for any $v \in V^1(\Omega)$.*

**Proof.** The estimate (6) is a direct consequence of the Poincaré inequality (see, e.g., [24], Chapter II, Theorem II.5.1). □

## 3. Strong Formulation of Problem (1) and Main Results

Let us suppose that

$$f \in C([0,+\infty);L^2(\Omega)), \qquad u_0 \in V^2(\Omega). \tag{7}$$

**Definition 1.** *We say that a pair $(u,\overline{p})$ is a strong solution to problem (1) if*

$$u \in C^1([0,+\infty);V^2(\Omega)), \qquad \overline{p} \in C([0,+\infty);H^1(\Omega)/\mathbb{R})$$

*and the following equalities are valid:*

$$u'(t) + (u(t) \cdot \nabla)u(t) - \nu\Delta u(t) - \alpha^2\Delta u'(t) + \nabla\overline{p}(t) = f(t), \quad t > 0, \qquad (8)$$
$$u(0) = u_0.$$

**Remark 1.** *Equation (8) with the initial condition $u(0) = u_0$ is a natural interpretation of the initial-boundary value problem (1) as an evolutionary equation in suitable function spaces. Note that, if a pair $(u_*, p_*)$ is a classical solution to problem (1), then $(u_*, \overline{p}_*)$ satisfies Equation (8), i.e., this pair is a strong solution. On the other hand, if $(u, \overline{p})$ is a strong solution and the functions $u$ and $p$ are sufficiently smooth in the usual sense, then $(u, p)$ is a classical solution to (1).*

We are now in a position to state our main results.

**Theorem 1.** *Assume that the boundary of the domain $\Omega$ belongs to the class $C^2$ and condition (7) holds. Then problem (1) has a unique strong solution $(u, \overline{p})$. This strong solution satisfies the energy equality*

$$\|u(t)\|^2_{L^2(\Omega)} + 2\nu \int_0^t \|\nabla u(\tau)\|^2_{L^2(\Omega)}\, d\tau + \alpha^2\|\nabla u(t)\|^2_{L^2(\Omega)}$$
$$= \|u_0\|^2_{L^2(\Omega)} + \alpha^2\|\nabla u_0\|^2_{L^2(\Omega)} + 2\int_0^t \int_\Omega f(\tau) \cdot u(\tau)\, dx\, d\tau, \quad t \geq 0. \qquad (9)$$

*If there exists a function $q \in C([0, T]; H^1(\Omega))$ such that $\nabla q = f$, then*

$$\|u(t)\|^2_{L^2(\Omega)} + \alpha^2\|\nabla u(t)\|^2_{L^2(\Omega)} \leq \exp\left(-\frac{8\nu t}{d^2_\Omega + 4\alpha^2}\right)\left(\|u_0\|^2_{L^2(\Omega)} + \alpha^2\|\nabla u_0\|^2_{L^2(\Omega)}\right), \quad t \geq 0, \qquad (10)$$

*where the positive constant $d_\Omega$ is defined in Lemma 3.*

## 4. Proof of Theorem 1

To prove the existence of a strong solution to problem (1), we use the Faedo–Galerkin method with a special basis of eigenfunctions of the Stokes operator

$$\mathbb{A}\colon V^2(\Omega) \to V^0(\Omega), \qquad \mathbb{A}w \stackrel{\text{def}}{=} -\mathbb{P}\Delta w.$$

This linear operator is invertible and $\mathbb{A}^{-1}$ is self-adjoint and compact as a map from $V^0(\Omega)$ into $V^0(\Omega)$. From the spectral theorem for self-adjoint compact operators (see, e.g., [25], Chapter 10, Theorem 10.12), it follows that there exist sequences $\{w_j\}_{j=1}^\infty \subset V^2(\Omega)$ and $\{\lambda_j\}_{j=1}^\infty \subset (0, +\infty)$ such that

$$\mathbb{A}w_j = \lambda_j w_j, \quad j \in \{1, 2, \ldots\}, \qquad (11)$$

and $\{w_j\}_{j=1}^\infty$ is an orthonormal basis of the space $V^0(\Omega)$.

Let

$$\widetilde{w}_j \stackrel{\text{def}}{=} \lambda_j^{-1} w_j, \quad j \in \{1, 2, \ldots\}.$$

It is easily shown that $\{\widetilde{w}_j\}_{j=1}^\infty$ is an orthonormal basis in the space $V^2(\Omega)$.

Let us fix an arbitrary number $T > 0$. For each fixed integer $m \geq 1$, we would like to define the approximate solution as follows:

$$v_m(t) \stackrel{\text{def}}{=} \sum_{i=1}^m g_{mi}(t) w_i, \quad t \in [0, T],$$

where $g_{m1}, \ldots, g_{mm}$ are unknown functions such that

$$\begin{cases} \int\limits_\Omega \boldsymbol{v}'_m(t) \cdot \boldsymbol{w}_j \, d\boldsymbol{x} + \sum\limits_{i=1}^n \int\limits_\Omega v_{mi}(t) \dfrac{\partial \boldsymbol{v}_m(t)}{\partial x_i} \cdot \boldsymbol{w}_j \, d\boldsymbol{x} - \nu \int\limits_\Omega \Delta \boldsymbol{v}_m(t) \cdot \boldsymbol{w}_j \, d\boldsymbol{x} \\ -\alpha^2 \int\limits_\Omega \Delta \boldsymbol{v}'_m(t) \cdot \boldsymbol{w}_j \, d\boldsymbol{x} = \int\limits_\Omega \boldsymbol{f}(t) \cdot \boldsymbol{w}_j \, d\boldsymbol{x}, \quad t \in (0,T), \ j = 1, \ldots, m, \\ \boldsymbol{v}_m(0) = \sum\limits_{i=1}^m (\boldsymbol{u}_0, \widetilde{\boldsymbol{w}}_i)_{\boldsymbol{V}^2(\Omega)} \widetilde{\boldsymbol{w}}_i. \end{cases} \tag{12}$$

Let us define the matrix $\boldsymbol{Q}_m \in \mathbb{R}^{m \times m}$ and the vector $\boldsymbol{a}_m \in \mathbb{R}^m$ by the rules:

$$Q_{mij} \stackrel{\text{def}}{=} \int\limits_\Omega \boldsymbol{w}_i \cdot \boldsymbol{w}_j \, d\boldsymbol{x} - \alpha^2 \int\limits_\Omega \Delta \boldsymbol{w}_i \cdot \boldsymbol{w}_j \, d\boldsymbol{x}, \quad i, j = 1, \ldots, m,$$

$$a_{mi} \stackrel{\text{def}}{=} \lambda_i^{-2} (\boldsymbol{u}_0, \boldsymbol{w}_i)_{\boldsymbol{V}^2(\Omega)}, \quad i = 1, \ldots, m.$$

Then, system (12) can be rewritten in the form

$$\begin{cases} \boldsymbol{Q}_m^\top \boldsymbol{g}'_m(t) = \mathbf{F}_m(t, \boldsymbol{g}_m(t)), \quad t \in (0,T), \\ \boldsymbol{g}_m(0) = \boldsymbol{a}_m, \end{cases} \tag{13}$$

where $\mathbf{F}_m \colon [0,T] \times \mathbb{R}^m \to \mathbb{R}^m$ is a known nonlinear vector function and $\boldsymbol{g}_m \stackrel{\text{def}}{=} (g_{m1}, \ldots, g_{mm})$.

Using integration by parts, we obtain

$$Q_{mij} = (\boldsymbol{w}_i, \boldsymbol{w}_j)_{\boldsymbol{V}^1(\Omega)}, \quad i, j = 1, \ldots, m.$$

Therefore, the matrix $\boldsymbol{Q}_m$ is symmetric and invertible.

Applying $\boldsymbol{Q}_m^{-1}$ to the first equation of problem (13), we obviously get

$$\begin{cases} \boldsymbol{g}'_m(t) = \boldsymbol{Q}_m^{-1} \mathbf{F}_m(t, \boldsymbol{g}_m(t)), \quad t \in (0,T), \\ \boldsymbol{g}_m(0) = \boldsymbol{a}_m. \end{cases}$$

The local existence of $\boldsymbol{g}_m$ on an interval $[0, T_m]$ is insured by the Cauchy–Peano theorem. Thus, we have a local solution $\boldsymbol{v}_m$ of problem (12) on $[0, T_m]$. Below, we obtain *a priori* estimates (independent of $m$) for vector function $\boldsymbol{v}_m$, which entail that $T_m = T$.

Let us assume that $\boldsymbol{v}_m$ satisfies system (12). We multiply the $j$th equation of (12) by $g_{mj}(t)$ and sum with respect to $j$ from 1 to $m$. Since

$$\begin{aligned} \sum_{i=1}^n \int\limits_\Omega v_{mi}(t) \frac{\partial \boldsymbol{v}_m(t)}{\partial x_i} \cdot \boldsymbol{v}_m(t) \, d\boldsymbol{x} &= \frac{1}{2} \sum_{i=1}^n \int\limits_\Omega v_{mi}(t) \frac{\partial}{\partial x_i} |\boldsymbol{v}_m(t)|^2 \, d\boldsymbol{x} \\ &= -\frac{1}{2} \sum_{i=1}^n \int\limits_\Omega \frac{\partial v_{mi}(t)}{\partial x_i} |\boldsymbol{v}_m(t)|^2 \, d\boldsymbol{x} \\ &= -\frac{1}{2} \int\limits_\Omega \underbrace{\nabla \cdot \boldsymbol{v}_m(t)}_{=0} |\boldsymbol{v}_m(t)|^2 \, d\boldsymbol{x} \\ &= 0, \quad t \in (0,T), \end{aligned}$$

we get

$$\int_\Omega \boldsymbol{v}'_m(t)\cdot\boldsymbol{v}_m(t)\,d\boldsymbol{x} - \nu\int_\Omega \Delta\boldsymbol{v}_m(t)\cdot\boldsymbol{v}_m(t)\,d\boldsymbol{x} - \alpha^2\int_\Omega \Delta\boldsymbol{v}'_m(t)\cdot\boldsymbol{v}_m(t)\,d\boldsymbol{x} = \int_\Omega \boldsymbol{f}(t)\cdot\boldsymbol{v}_m(t)\,d\boldsymbol{x}. \quad (14)$$

Integrating by parts the second and third terms on the left-hand side of equality (14), we arrive at the following relation

$$\int_\Omega \boldsymbol{v}'_m(t)\cdot\boldsymbol{v}_m(t)\,d\boldsymbol{x} + \nu\int_\Omega |\nabla\boldsymbol{v}_m(t)|^2\,d\boldsymbol{x} + \alpha^2\int_\Omega \nabla\boldsymbol{v}'_m(t):\nabla\boldsymbol{v}_m(t)\,d\boldsymbol{x} = \int_\Omega \boldsymbol{f}(t)\cdot\boldsymbol{v}_m(t)\,d\boldsymbol{x},$$

which, in turn, gives

$$\frac{1}{2}\frac{d}{d\tau}\|\boldsymbol{v}_m(\tau)\|^2_{\boldsymbol{L}^2(\Omega)} + \nu\|\nabla\boldsymbol{v}_m(\tau)\|^2_{\boldsymbol{L}^2(\Omega)} + \frac{\alpha^2}{2}\frac{d}{d\tau}\|\nabla\boldsymbol{v}_m(\tau)\|^2_{\boldsymbol{L}^2(\Omega)} = \int_\Omega \boldsymbol{f}(\tau)\cdot\boldsymbol{v}_m(\tau)\,d\boldsymbol{x},$$

for any $\tau\in[0,T]$. Further, we multiply the last equality by 2 and integrate from 0 to $t$ with respect to $\tau$; this yields

$$\begin{aligned}\|\boldsymbol{v}_m(t)\|^2_{\boldsymbol{L}^2(\Omega)} + 2\nu\int_0^t \|\nabla\boldsymbol{v}_m(\tau)\|^2_{\boldsymbol{L}^2(\Omega)}\,d\tau + \alpha^2\|\nabla\boldsymbol{v}_m(t)\|^2_{\boldsymbol{L}^2(\Omega)}&\\ = \|\boldsymbol{v}_m(0)\|^2_{\boldsymbol{L}^2(\Omega)} + \alpha^2\|\nabla\boldsymbol{v}_m(0)\|^2_{\boldsymbol{L}^2(\Omega)} + 2\int_0^t\int_\Omega \boldsymbol{f}(\tau)\cdot\boldsymbol{v}_m(\tau)\,d\boldsymbol{x}\,d\tau.&\quad (15)\end{aligned}$$

Taking into account (3) and (4), we easily derive from equality (15) that

$$\begin{aligned}\|\boldsymbol{v}_m(t)\|^2_{\boldsymbol{V}^1(\Omega)} &\le \|\boldsymbol{v}_m(0)\|^2_{\boldsymbol{V}^1(\Omega)} + 2\int_0^t\int_\Omega \boldsymbol{f}(\tau)\cdot\boldsymbol{v}_m(\tau)\,d\boldsymbol{x}\,d\tau\\ &\le \|\boldsymbol{v}_m(0)\|^2_{\boldsymbol{V}^1(\Omega)} + \int_0^t\int_\Omega |\boldsymbol{f}(\tau)|^2\,d\boldsymbol{x}\,d\tau + \int_0^t\int_\Omega |\boldsymbol{v}_m(\tau)|^2\,d\boldsymbol{x}\,d\tau\\ &\le C_1\|\boldsymbol{u}_0\|^2_{\boldsymbol{V}^2(\Omega)} + \int_0^T \|\boldsymbol{f}(\tau)\|^2_{\boldsymbol{L}^2(\Omega)}\,d\tau + C_2\int_0^t \|\boldsymbol{v}_m(\tau)\|^2_{\boldsymbol{V}^1(\Omega)}\,d\tau.\end{aligned}$$

Here and in the succeeding discussion, the symbols $C_i$, $i=1,2,\ldots$, designate positive constants that are independent of $m$. Using Grönwall's inequality, we get

$$\|\boldsymbol{v}_m(t)\|^2_{\boldsymbol{V}^1(\Omega)} \le \left(C_1\|\boldsymbol{u}_0\|^2_{\boldsymbol{V}^2(\Omega)} + \int_0^T \|\boldsymbol{f}(\tau)\|^2_{\boldsymbol{L}^2(\Omega)}\,d\tau\right)\exp(C_2t), \quad t\in(0,T). \quad (16)$$

Hence,

$$\begin{aligned}\|\boldsymbol{v}_m\|_{\boldsymbol{C}([0,T];\boldsymbol{V}^1(\Omega))} &= \max_{t\in[0,T]}\|\boldsymbol{v}_m(t)\|_{\boldsymbol{V}^1(\Omega)}\\ &\le \left(C_1\|\boldsymbol{u}_0\|^2_{\boldsymbol{V}^2(\Omega)} + \int_0^T \|\boldsymbol{f}(\tau)\|^2_{\boldsymbol{L}^2(\Omega)}\,d\tau\right)^{1/2}\left[\exp(C_2T)\right]^{1/2}. \quad (17)\end{aligned}$$

Next, by multiplying the $j$th equation of (12) with $g'_{mj}$ and summing over $j = 1, \dots, m$, we obtain

$$\int_\Omega |\boldsymbol{v}'_m(t)|^2\, \boldsymbol{dx} + \sum_{i=1}^n \int_\Omega v_{mi}(t) \frac{\partial \boldsymbol{v}_m(t)}{\partial x_i} \cdot \boldsymbol{v}'_m(t)\, \boldsymbol{dx} - \nu \int_\Omega \Delta \boldsymbol{v}_m(t) \cdot \boldsymbol{v}'_m(t)\, \boldsymbol{dx}$$
$$- \alpha^2 \int_\Omega \Delta \boldsymbol{v}'_m(t) \cdot \boldsymbol{v}'_m(t)\, \boldsymbol{dx} = \int_\Omega \boldsymbol{f}(t) \cdot \boldsymbol{v}'_m(t)\, \boldsymbol{dx}, \quad t \in (0,T).$$

Integrating by parts the third and fourth terms on the left-hand side of the last equality, we arrive at

$$\int_\Omega |\boldsymbol{v}'_m(t)|^2\, \boldsymbol{dx} + \sum_{i=1}^n \int_\Omega v_{mi}(t) \frac{\partial \boldsymbol{v}_m(t)}{\partial x_i} \cdot \boldsymbol{v}'_m(t)\, \boldsymbol{dx} + \nu \int_\Omega \nabla \boldsymbol{v}_m(t) : \nabla \boldsymbol{v}'_m(t)\, \boldsymbol{dx}$$
$$+ \alpha^2 \int_\Omega |\nabla \boldsymbol{v}'_m(t)|^2\, \boldsymbol{dx} = \int_\Omega \boldsymbol{f}(t) \cdot \boldsymbol{v}'_m(t)\, \boldsymbol{dx}, \quad t \in (0,T).$$

From here, using (3) and Hölder's inequality, one can obtain

$$\begin{aligned} \|\boldsymbol{v}'_m(t)\|^2_{\boldsymbol{V}^1(\Omega)} &= - \sum_{i=1}^n \int_\Omega v_{mi}(t) \frac{\partial \boldsymbol{v}_m(t)}{\partial x_i} \cdot \boldsymbol{v}'_m(t)\, \boldsymbol{dx} \\ &\quad - \nu \int_\Omega \nabla \boldsymbol{v}_m(t) : \nabla \boldsymbol{v}'_m(t)\, \boldsymbol{dx} + \int_\Omega \boldsymbol{f}(t) \cdot \boldsymbol{v}'_m(t)\, \boldsymbol{dx} \\ &\le \sum_{i,j=1}^n \|v_{mi}(t)\|_{L^4(\Omega)} \left\| \frac{\partial v_{mj}(t)}{\partial x_i} \right\|_{L^2(\Omega)} \|v'_{mj}(t)\|_{L^4(\Omega)} \\ &\quad + \nu \|\nabla \boldsymbol{v}_m(t)\|_{\boldsymbol{L}^2(\Omega)} \|\nabla \boldsymbol{v}'_m(t)\|_{\boldsymbol{L}^2(\Omega)} + \|\boldsymbol{f}(t)\|_{\boldsymbol{L}^2(\Omega)} \|\boldsymbol{v}'_m(t)\|_{\boldsymbol{L}^2(\Omega)} \\ &\le C_3 \big( \|\boldsymbol{v}_m(t)\|^2_{\boldsymbol{V}^1(\Omega)} + \|\boldsymbol{f}(t)\|_{\boldsymbol{L}^2(\Omega)} \big) \|\boldsymbol{v}'_m(t)\|_{\boldsymbol{V}^1(\Omega)}, \end{aligned}$$

whence

$$\|\boldsymbol{v}'_m(t)\|_{\boldsymbol{V}^1(\Omega)} \le C_3 \big( \|\boldsymbol{v}_m(t)\|^2_{\boldsymbol{V}^1(\Omega)} + \|\boldsymbol{f}(t)\|_{\boldsymbol{L}^2(\Omega)} \big), \quad t \in (0,T).$$

With the help of inequality (16), we get

$$\|\boldsymbol{v}'_m(t)\|_{\boldsymbol{V}^1(\Omega)} \le C_3 \left( C_1 \|\boldsymbol{u}_0\|^2_{\boldsymbol{V}^2(\Omega)} + \int_0^T \|\boldsymbol{f}(\tau)\|^2_{\boldsymbol{L}^2(\Omega)}\, d\tau \right) \exp(C_2 t) + C_3 \|\boldsymbol{f}(t)\|_{\boldsymbol{L}^2(\Omega)},$$

for all $t \in (0,T)$. Therefore, we have

$$\begin{aligned} \|\boldsymbol{v}'_m\|_{C([0,T];\boldsymbol{V}^1(\Omega))} &= \max_{t \in [0,T]} \|\boldsymbol{v}'_m(t)\|_{\boldsymbol{V}^1(\Omega)} \\ &\le C_3 \left( C_1 \|\boldsymbol{u}_0\|^2_{\boldsymbol{V}^2(\Omega)} + \int_0^T \|\boldsymbol{f}(\tau)\|^2_{\boldsymbol{L}^2(\Omega)}\, d\tau \right) \exp(C_2 T) + C_3 \max_{t \in [0,T]} \|\boldsymbol{f}(t)\|_{\boldsymbol{L}^2(\Omega)}. \end{aligned} \tag{18}$$

We now multiply the $j$th equation of (12) by $-\lambda_j g_{mj}(t)$ and sum with respect to $j$ from 1 to $m$. Taking into account equality (11), we get

$$\int_\Omega \boldsymbol{v}'_m(t) \cdot \mathbb{P}\Delta \boldsymbol{v}_m(t)\, \boldsymbol{dx} + \sum_{i=1}^n \int_\Omega v_{mi}(t) \frac{\partial \boldsymbol{v}_m(t)}{\partial x_i} \cdot \mathbb{P}\Delta \boldsymbol{v}_m(t)\, \boldsymbol{dx} - \nu \int_\Omega \Delta \boldsymbol{v}_m(t) \cdot \mathbb{P}\Delta \boldsymbol{v}_m(t)\, \boldsymbol{dx}$$
$$- \alpha^2 \int_\Omega \Delta \boldsymbol{v}'_m(t) \cdot \mathbb{P}\Delta \boldsymbol{v}_m(t)\, \boldsymbol{dx} = \int_\Omega \boldsymbol{f}(t) \cdot \mathbb{P}\Delta \boldsymbol{v}_m(t)\, \boldsymbol{dx}, \quad t \in (0,T),$$

which leads to

$$\int_\Omega \boldsymbol{v}'_m(t) \cdot \mathbb{P}\Delta \boldsymbol{v}_m(t)\, \boldsymbol{dx} + \sum_{i=1}^n \int_\Omega v_{mi}(t) \frac{\partial \boldsymbol{v}_m(t)}{\partial x_i} \cdot \mathbb{P}\Delta \boldsymbol{v}_m(t)\, \boldsymbol{dx} - \nu \int_\Omega |\mathbb{P}\Delta \boldsymbol{v}_m(t)|^2\, \boldsymbol{dx}$$
$$- \alpha^2 \int_\Omega \mathbb{P}\Delta \boldsymbol{v}'_m(t) \cdot \mathbb{P}\Delta \boldsymbol{v}_m(t)\, \boldsymbol{dx} = \int_\Omega \boldsymbol{f}(t) \cdot \mathbb{P}\Delta \boldsymbol{v}_m(t)\, \boldsymbol{dx}, \quad t \in (0,T).$$

From this equality, with the help of Hölder's and Young's inequalities, we derive

$$\begin{aligned}
&\nu \|\mathbb{P}\Delta \boldsymbol{v}_m(t)\|^2_{\boldsymbol{L}^2(\Omega)} + \frac{\alpha^2}{2} \frac{d}{dt} \|\mathbb{P}\Delta \boldsymbol{v}_m(t)\|^2_{\boldsymbol{L}^2(\Omega)} \\
&= \int_\Omega \boldsymbol{v}'_m(t) \cdot \mathbb{P}\Delta \boldsymbol{v}_m(t)\, \boldsymbol{dx} + \sum_{i=1}^n \int_\Omega v_{mi}(t) \frac{\partial \boldsymbol{v}_m(t)}{\partial x_i} \cdot \mathbb{P}\Delta \boldsymbol{v}_m(t)\, \boldsymbol{dx} - \int_\Omega \boldsymbol{f}(t) \cdot \mathbb{P}\Delta \boldsymbol{v}_m(t)\, \boldsymbol{dx} \\
&\leq \Big( \|\boldsymbol{v}'_m(t)\|_{\boldsymbol{L}^2(\Omega)} + \sum_{i=1}^n \|v_{mi}(t)\|_{L^4(\Omega)} \Big\| \frac{\partial \boldsymbol{v}_m(t)}{\partial x_i} \Big\|_{\boldsymbol{L}^4(\Omega)} + \|\boldsymbol{f}(t)\|_{\boldsymbol{L}^2(\Omega)} \Big) \|\mathbb{P}\Delta \boldsymbol{v}_m(t)\|_{\boldsymbol{L}^2(\Omega)} \\
&\leq \frac{1}{2\nu} \Big( \|\boldsymbol{v}'_m(t)\|_{\boldsymbol{L}^2(\Omega)} + \sum_{i=1}^n \|v_{mi}(t)\|_{L^4(\Omega)} \Big\| \frac{\partial \boldsymbol{v}_m(t)}{\partial x_i} \Big\|_{\boldsymbol{L}^4(\Omega)} + \|\boldsymbol{f}(t)\|_{\boldsymbol{L}^2(\Omega)} \Big)^2 \\
&\quad + \frac{\nu}{2} \|\mathbb{P}\Delta \boldsymbol{v}_m(t)\|^2_{\boldsymbol{L}^2(\Omega)}, \quad t \in (0,T).
\end{aligned}$$

Therefore, the following inequality holds

$$\begin{aligned}
&\nu \|\mathbb{P}\Delta \boldsymbol{v}_m(t)\|^2_{\boldsymbol{L}^2(\Omega)} + \alpha^2 \frac{d}{dt} \|\mathbb{P}\Delta \boldsymbol{v}_m(t)\|^2_{\boldsymbol{L}^2(\Omega)} \\
&\quad \leq \frac{1}{\nu} \Big( \|\boldsymbol{v}'_m(t)\|_{\boldsymbol{L}^2(\Omega)} + \sum_{i=1}^n \|v_{mi}(t)\|_{L^4(\Omega)} \Big\| \frac{\partial \boldsymbol{v}_m(t)}{\partial x_i} \Big\|_{\boldsymbol{L}^4(\Omega)} + \|\boldsymbol{f}(t)\|_{\boldsymbol{L}^2(\Omega)} \Big)^2, \quad t \in (0,T),
\end{aligned}$$

and, using estimates (17) and (18), we deduce that

$$\nu \|\boldsymbol{v}_m(\tau)\|^2_{\boldsymbol{V}^2(\Omega)} + \alpha^2 \frac{d}{d\tau} \|\boldsymbol{v}_m(\tau)\|^2_{\boldsymbol{V}^2(\Omega)} \leq C_4 + C_5 \|\boldsymbol{v}_m(\tau)\|^2_{\boldsymbol{V}^2(\Omega)}, \quad \tau \in (0,T).$$

Integrating both sides of this differential inequality with respect to $\tau$ from 0 to $t$, we deduce

$$\begin{aligned}
\nu \int_0^t \|\boldsymbol{v}_m(\tau)\|^2_{\boldsymbol{V}^2(\Omega)}\, d\tau + \alpha^2 \|\boldsymbol{v}_m(t)\|^2_{\boldsymbol{V}^2(\Omega)} &\leq \alpha^2 \|\boldsymbol{v}_m(0)\|^2_{\boldsymbol{V}^2(\Omega)} + C_4 t + C_5 \int_0^t \|\boldsymbol{v}_m(\tau)\|^2_{\boldsymbol{V}^2(\Omega)} d\tau \\
&\leq \alpha^2 \|\boldsymbol{u}_0\|^2_{\boldsymbol{V}^2(\Omega)} + C_4 T + C_5 \int_0^t \|\boldsymbol{v}_m(\tau)\|^2_{\boldsymbol{V}^2(\Omega)}\, d\tau.
\end{aligned}$$

It follows easily that

$$\|\boldsymbol{v}_m(t)\|^2_{\boldsymbol{V}^2(\Omega)} \leq \|\boldsymbol{u}_0\|^2_{\boldsymbol{V}^2(\Omega)} + C_4 \alpha^{-2} T + C_5 \alpha^{-2} \int_0^t \|\boldsymbol{v}_m(\tau)\|^2_{\boldsymbol{V}^2(\Omega)}\, d\tau, \quad t \in (0,T).$$

Applying Grönwall's inequality, we obtain

$$\|\boldsymbol{v}_m(t)\|^2_{\boldsymbol{V}^2(\Omega)} \le \Big(\|\boldsymbol{u}_0\|^2_{\boldsymbol{V}^2(\Omega)} + C_4\alpha^{-2}T\Big)\exp(C_5\alpha^{-2}t), \quad t\in(0,T).$$

This implies that

$$\|\boldsymbol{v}_m\|_{\boldsymbol{C}([0,T];\boldsymbol{V}^2(\Omega))} = \max_{t\in[0,T]}\|\boldsymbol{v}_m(t)\|_{\boldsymbol{V}^2(\Omega)} \le \Big\{\big(\|\boldsymbol{u}_0\|^2_{\boldsymbol{V}^2(\Omega)} + C_4\alpha^{-2}T\big)\exp(C_5\alpha^{-2}T)\Big\}^{1/2}. \tag{19}$$

Finally, we multiply the $j$th equation of (12) by $-\lambda_j g'_{mj}(t)$ and sum with respect to $j$ from 1 to $m$. Bearing in mind equality (11), we obtain

$$\int_\Omega \boldsymbol{v}'_m(t)\cdot\mathbb{P}\Delta\boldsymbol{v}'_m(t)\,\boldsymbol{dx} + \sum_{i=1}^n\int_\Omega v_{mi}(t)\frac{\partial\boldsymbol{v}_m(t)}{\partial x_i}\cdot\mathbb{P}\Delta\boldsymbol{v}'_m(t)\,\boldsymbol{dx} - \nu\int_\Omega\Delta\boldsymbol{v}_m(t)\cdot\mathbb{P}\Delta\boldsymbol{v}'_m(t)\,\boldsymbol{dx}$$
$$-\alpha^2\int_\Omega\Delta\boldsymbol{v}'_m(t)\cdot\mathbb{P}\Delta\boldsymbol{v}'_m(t)\,\boldsymbol{dx} = \int_\Omega\boldsymbol{f}(t)\cdot\mathbb{P}\Delta\boldsymbol{v}'_m(t)\,\boldsymbol{dx}, \quad t\in(0,T).$$

Using Hölder's inequality, from the last equality one can derive

$$\begin{aligned}
\alpha^2\|\mathbb{P}\Delta\boldsymbol{v}'_m(t)\|^2_{\boldsymbol{L}^2(\Omega)} =& \int_\Omega \boldsymbol{v}'_m(t)\cdot\mathbb{P}\Delta\boldsymbol{v}'_m(t)\,\boldsymbol{dx} + \sum_{i=1}^n\int_\Omega v_{mi}(t)\frac{\partial\boldsymbol{v}_m(t)}{\partial x_i}\cdot\mathbb{P}\Delta\boldsymbol{v}'_m(t)\,\boldsymbol{dx}\\
&-\nu\int_\Omega\mathbb{P}\Delta\boldsymbol{v}_m(t)\cdot\mathbb{P}\Delta\boldsymbol{v}'_m(t)\,\boldsymbol{dx} - \int_\Omega\boldsymbol{f}(t)\cdot\mathbb{P}\Delta\boldsymbol{v}'_m(t)\,\boldsymbol{dx}\\
\le&\Big(\|\boldsymbol{v}'_m(t)\|_{\boldsymbol{L}^2(\Omega)} + \sum_{i=1}^n\|v_{mi}(t)\|_{L^4(\Omega)}\Big\|\frac{\partial\boldsymbol{v}_m(t)}{\partial x_i}\Big\|_{\boldsymbol{L}^4(\Omega)}\\
&+\nu\|\mathbb{P}\Delta\boldsymbol{v}_m(t)\|_{\boldsymbol{L}^2(\Omega)} + \|\boldsymbol{f}(t)\|_{\boldsymbol{L}^2(\Omega)}\Big)\|\mathbb{P}\Delta\boldsymbol{v}'_m(t)\|_{\boldsymbol{L}^2(\Omega)}\\
\le&\Big(\|\boldsymbol{v}'_m(t)\|_{\boldsymbol{L}^2(\Omega)} + C_6\|\boldsymbol{v}_m(t)\|_{\boldsymbol{V}^1(\Omega)}\|\boldsymbol{v}_m(t)\|_{\boldsymbol{V}^2(\Omega)}\\
&+\nu\|\boldsymbol{v}_m(t)\|_{\boldsymbol{V}^2(\Omega)} + \|\boldsymbol{f}(t)\|_{\boldsymbol{L}^2(\Omega)}\Big)\|\mathbb{P}\Delta\boldsymbol{v}'_m(t)\|_{\boldsymbol{L}^2(\Omega)}, \quad t\in(0,T).
\end{aligned}$$

Clearly, this yields the estimate

$$\begin{aligned}
\|\mathbb{P}\Delta\boldsymbol{v}'_m(t)\|_{\boldsymbol{L}^2(\Omega)} \le& \alpha^{-2}\Big(\|\boldsymbol{v}'_m(t)\|_{\boldsymbol{L}^2(\Omega)} + C_6\|\boldsymbol{v}_m(t)\|_{\boldsymbol{V}^1(\Omega)}\|\boldsymbol{v}_m(t)\|_{\boldsymbol{V}^2(\Omega)}\\
&+\nu\|\boldsymbol{v}_m(t)\|_{\boldsymbol{V}^2(\Omega)} + \|\boldsymbol{f}(t)\|_{\boldsymbol{L}^2(\Omega)}\Big), \quad t\in(0,T).
\end{aligned}$$

Taking into account (17)–(19), from the last inequality, we easily obtain that

$$\|\boldsymbol{v}'_m(t)\|_{\boldsymbol{V}^2(\Omega)} = \|\mathbb{P}\Delta\boldsymbol{v}'_m(t)\|_{\boldsymbol{L}^2(\Omega)} \le C_7, \quad t\in(0,T),$$

and, hence,

$$\|\boldsymbol{v}'_m\|_{\boldsymbol{C}([0,T];\boldsymbol{V}^2(\Omega))} = \max_{t\in[0,T]}\|\boldsymbol{v}'_m(t)\|_{\boldsymbol{V}^2(\Omega)} \le C_7. \tag{20}$$

From estimates (19) and (20) and Lemma 2, it follows that there exist a subsequence $\{m_k\}_{k=1}^\infty$ and a function $\boldsymbol{u}$ such that $\boldsymbol{v}_{m_k}$ converges strongly to $\boldsymbol{u}$ in the space $\boldsymbol{C}([0,T];\boldsymbol{V}^1(\Omega))$ as $k\to\infty$. Without loss of generality, we can assume that

$$\boldsymbol{v}_m \to \boldsymbol{u} \ \text{ strongly in } \boldsymbol{C}([0,T];\boldsymbol{V}^1(\Omega)) \text{ as } m \to \infty, \tag{21}$$

$$\boldsymbol{v}_m \rightharpoonup \boldsymbol{u} \ \text{ weakly in } \boldsymbol{L}^2(0,T;\boldsymbol{V}^2(\Omega)) \text{ as } m \to \infty. \tag{22}$$

Moreover, we have

$$\boldsymbol{v}_m(0) \to \boldsymbol{u}_0 \ \text{ strongly in } \boldsymbol{V}^2(\Omega) \text{ as } m \to \infty. \tag{23}$$

On the other hand, from (21) it follows that

$$\boldsymbol{v}_m(0) \to \boldsymbol{u}(0) \ \text{ strongly in } \boldsymbol{V}^1(\Omega) \text{ as } m \to \infty. \tag{24}$$

Comparing the convergence results (23) and (24), we obtain

$$\boldsymbol{u}(0) = \boldsymbol{u}_0. \tag{25}$$

Integrating the $j$th equation of (12) from 0 to $s$, we obtain

$$\int_\Omega \boldsymbol{v}_m(s)\cdot\boldsymbol{w}_j\,\boldsymbol{dx} + \sum_{i=1}^n \int_0^s\int_\Omega v_{mi}(t)\frac{\partial\boldsymbol{v}_m(t)}{\partial x_i}\cdot\boldsymbol{w}_j\,\boldsymbol{dx}dt - \nu\int_0^s\int_\Omega \Delta\boldsymbol{v}_m(t)\cdot\boldsymbol{w}_j\,\boldsymbol{dx}dt - \alpha^2\int_\Omega \Delta\boldsymbol{v}_m(s)\cdot\boldsymbol{w}_j\,\boldsymbol{dx}$$
$$= \int_\Omega \boldsymbol{v}_m(0)\cdot\boldsymbol{w}_j\,\boldsymbol{dx} - \alpha^2\int_\Omega \Delta\boldsymbol{v}_m(0)\cdot\boldsymbol{w}_j\,\boldsymbol{dx} + \int_0^s\int_\Omega \boldsymbol{f}(t)\cdot\boldsymbol{w}_j\,\boldsymbol{dx}dt, \quad j\in\{1,2,\dots\},\ s\in[0,T].$$

Integrating by parts the third and fourth terms on the left-hand side of this equality, we arrive at

$$\int_\Omega \boldsymbol{v}_m(s)\cdot\boldsymbol{w}_j\,\boldsymbol{dx} + \sum_{i=1}^n \int_0^s\int_\Omega v_{mi}(t)\frac{\partial\boldsymbol{v}_m(t)}{\partial x_i}\cdot\boldsymbol{w}_j\,\boldsymbol{dx}dt + \nu\int_0^s\int_\Omega \nabla\boldsymbol{v}_m(t):\nabla\boldsymbol{w}_j\,\boldsymbol{dx}dt$$
$$+\alpha^2\int_\Omega \nabla\boldsymbol{v}_m(s):\nabla\boldsymbol{w}_j\,\boldsymbol{dx} = \int_\Omega \boldsymbol{v}_m(0)\cdot\boldsymbol{w}_j\,\boldsymbol{dx} - \alpha^2\int_\Omega \Delta\boldsymbol{v}_m(0)\cdot\boldsymbol{w}_j\,\boldsymbol{dx} + \int_0^s\int_\Omega \boldsymbol{f}(t)\cdot\boldsymbol{w}_j\,\boldsymbol{dx}dt.$$

Using the convergence results (21)–(23), we can pass to the limit $m\to\infty$ in the last equality and obtain

$$\int_\Omega \boldsymbol{u}(s)\cdot\boldsymbol{w}_j\,\boldsymbol{dx} + \sum_{i=1}^n \int_0^s\int_\Omega u_i(t)\frac{\partial\boldsymbol{u}(t)}{\partial x_i}\cdot\boldsymbol{w}_j\,\boldsymbol{dx}dt + \nu\int_0^s\int_\Omega \nabla\boldsymbol{u}(t):\nabla\boldsymbol{w}_j\,\boldsymbol{dx}dt + \alpha^2\int_\Omega \nabla\boldsymbol{u}(s):\nabla\boldsymbol{w}_j\,\boldsymbol{dx}$$
$$= \int_\Omega \boldsymbol{u}_0\cdot\boldsymbol{w}_j\,\boldsymbol{dx} - \alpha^2\int_\Omega \Delta\boldsymbol{u}_0\cdot\boldsymbol{w}_j\,\boldsymbol{dx} + \int_0^s\int_\Omega \boldsymbol{f}(t)\cdot\boldsymbol{w}_j\,\boldsymbol{dx}dt, \quad j\in\{1,2,\dots\},\ s\in[0,T].$$

Applying integration by parts again, we get

$$\int_\Omega \boldsymbol{u}(s)\cdot\boldsymbol{w}_j\,\boldsymbol{dx} + \sum_{i=1}^n \int_0^s\int_\Omega u_i(t)\frac{\partial\boldsymbol{u}(t)}{\partial x_i}\cdot\boldsymbol{w}_j\,\boldsymbol{dx}dt - \nu\int_0^s\int_\Omega \Delta\boldsymbol{u}(t)\cdot\boldsymbol{w}_j\,\boldsymbol{dx}dt - \alpha^2\int_\Omega \Delta\boldsymbol{u}(s)\cdot\boldsymbol{w}_j\,\boldsymbol{dx}$$
$$= \int_\Omega \boldsymbol{u}_0\cdot\boldsymbol{w}_j\,\boldsymbol{dx} - \alpha^2\int_\Omega \Delta\boldsymbol{u}_0\cdot\boldsymbol{w}_j\,\boldsymbol{dx} + \int_0^s\int_\Omega \boldsymbol{f}(t)\cdot\boldsymbol{w}_j\,\boldsymbol{dx}dt, \quad j\in\{1,2,\dots\},\ s\in[0,T]. \tag{26}$$

Because $\{w_j\}_{j=1}^{\infty}$ is a basis of $V^0(\Omega)$, equality (26) remains valid if we replace $w_j$ with an arbitrary vector function $w$ from the space $V^0(\Omega)$, that is

$$\int_\Omega u(s)\cdot w\,dx+\sum_{i=1}^n\int_0^s\int_\Omega u_i(t)\frac{\partial u(t)}{\partial x_i}\cdot w\,dxdt-\nu\int_0^s\int_\Omega \Delta u(t)\cdot w\,dxdt$$
$$-\alpha^2\int_\Omega \Delta u(s)\cdot w\,dx=\int_\Omega u_0\cdot w\,dx-\alpha^2\int_\Omega \Delta u_0\cdot w\,dx+\int_0^s\int_\Omega f(t)\cdot w\,dxdt,\quad s\in[0,T].$$

From the last equality it follows that

$$u(s)+\sum_{i=1}^n\mathbb{P}\int_0^s u_i(t)\frac{\partial u(t)}{\partial x_i}dt-\nu\mathbb{P}\int_0^s\Delta u(t)dt-\alpha^2\mathbb{P}\Delta u(s)=u_0-\alpha^2\mathbb{P}\Delta u_0+\mathbb{P}\int_0^s f(t)dt. \quad (27)$$

Using the Stokes operator $\mathbb{A}$, we can rewrite this equality as follows

$$(\mathbb{I}+\alpha^2\mathbb{A})u(s)=-\sum_{i=1}^n\mathbb{P}\int_0^s u_i(t)\frac{\partial u(t)}{\partial x_i}dt+\nu\mathbb{P}\int_0^s\Delta u(t)dt+(\mathbb{I}+\alpha^2\mathbb{A})u_0+\mathbb{P}\int_0^s f(t)dt,\quad s\in[0,T], \quad (28)$$

where $\mathbb{I}\colon V^2(\Omega)\to V^0(\Omega)$ is the embedding operator.

Applying the operator $(\mathbb{I}+\alpha^2\mathbb{A})^{-1}\colon V^0(\Omega)\to V^2(\Omega)$ to both sides of equality (28), we get

$$u(s)=-\sum_{i=1}^n(\mathbb{I}+\alpha^2\mathbb{A})^{-1}\mathbb{P}\int_0^s u_i(t)\frac{\partial u(t)}{\partial x_i}dt+\nu(\mathbb{I}+\alpha^2\mathbb{A})^{-1}\mathbb{P}\int_0^s\Delta u(t)dt$$
$$+u_0+(\mathbb{I}+\alpha^2\mathbb{A})^{-1}\mathbb{P}\int_0^s f(t)dt,\quad s\in[0,T]. \quad (29)$$

Since

$$u\in C([0,T];V^1(\Omega))\cap L^2(0,T;V^2(\Omega)),$$

we conclude from (29) that

$$u\in C([0,T];V^2(\Omega)). \quad (30)$$

Next, differentiating both sides of (29) with respect to $s$, we get

$$u'(s)=-\sum_{i=1}^n(\mathbb{I}+\alpha^2\mathbb{A})^{-1}\mathbb{P}\left[u_i(s)\frac{\partial u(s)}{\partial x_i}\right]+\nu(\mathbb{I}+\alpha^2\mathbb{A})^{-1}\mathbb{P}\Delta u(s)+(\mathbb{I}+\alpha^2\mathbb{A})^{-1}\mathbb{P}f(s),\quad s\in[0,T].$$

Taking into account (30), from the last equality we deduce that $u'\in C([0,T];V^2(\Omega))$. Hence,

$$u\in C^1([0,T];V^2(\Omega)). \quad (31)$$

Next, from equality (27) it follows that there exists an element $\overline{\pi}(t)\in H^1(\Omega)/\mathbb{R}$ such that

$$u(t)+\sum_{i=1}^n\int_0^t u_i(\tau)\frac{\partial u(\tau)}{\partial x_i}\,d\tau-\nu\int_0^t\Delta u(\tau)\,d\tau-\alpha^2\Delta u(t)-u_0-\alpha^2\Delta u_0-\int_0^t f(\tau)\,d\tau=\nabla\overline{\pi}(t). \quad (32)$$

It is readily seen that $\nabla\overline{\pi}\in C^1([0,T];G(\Omega))$ and, consequently, we have

$$\overline{\pi}\in C^1([0,T];H^1(\Omega)/\mathbb{R}). \quad (33)$$

Letting $\overline{p}(t) \stackrel{\text{def}}{=} -\overline{\pi}'(t)$, from (33) we get

$$\overline{p} \in C([0,T]; H^1(\Omega)/\mathbb{R}). \tag{34}$$

Finally, differentiating both sides of (32) with respect to $t$, we arrive at

$$\boldsymbol{u}'(t) + \sum_{i=1}^{n} u_i(t) \frac{\partial \boldsymbol{u}(t)}{\partial x_i} - \nu \Delta \boldsymbol{u}(t) - \alpha^2 \Delta \boldsymbol{u}'(t) + \nabla \overline{p}(t) = \boldsymbol{f}(t), \quad t \in (0,T). \tag{35}$$

Bearing in mind (25), (31), (34), and (35), we conclude that the pair $(\boldsymbol{u}, \overline{p})$ is a strong solution to problem (1) on the interval $[0,T]$. The uniqueness of a strong solution can be proved by using arguments similar to those that are presented in [9], thus we choose to omit the details of the corresponding proof. Since $T$ is arbitrary, we see that $(\boldsymbol{u}, \overline{p})$ is a solution of (1) in the sense of Definition 1.

Next, we take the $L^2$-scalar product of (8) with the vector function $\boldsymbol{u}$. Using integration by parts, one can easily arrive at the energy equality (9).

The rest of the proof consists in proving inequality (10). If there exists a function $q$ from the space $C([0,T]; H^1(\Omega))$ such that $\nabla q = \boldsymbol{f}$, then we have

$$\int_\Omega \boldsymbol{f}(\tau) \cdot \boldsymbol{u}(\tau)\, d\boldsymbol{x} = \int_\Omega \nabla q(\tau) \cdot \boldsymbol{u}(\tau)\, d\boldsymbol{x} = -\int_\Omega q(\tau) \underbrace{\nabla \cdot \boldsymbol{u}(\tau)}_{=0}\, d\boldsymbol{x} = 0, \quad \tau \geq 0, \tag{36}$$

i. e., the total work done by external forces $\boldsymbol{f}$ is zero.

In view of (36), the energy equality (9) reduces to

$$\|\boldsymbol{u}(t)\|^2_{\boldsymbol{L}^2(\Omega)} + 2\nu \int_0^t \|\nabla \boldsymbol{u}(\tau)\|^2_{\boldsymbol{L}^2(\Omega)}\, d\tau + \alpha^2 \|\nabla \boldsymbol{u}(t)\|^2_{\boldsymbol{L}^2(\Omega)} = \|\boldsymbol{u}_0\|^2_{\boldsymbol{L}^2(\Omega)} + \alpha^2 \|\nabla \boldsymbol{u}_0\|^2_{\boldsymbol{L}^2(\Omega)}, \quad t \geq 0.$$

Differentiating the last equality with respect to $t$, we get

$$\frac{d}{dt}\left[\|\boldsymbol{u}(t)\|^2_{\boldsymbol{L}^2(\Omega)} + \alpha^2 \|\nabla \boldsymbol{u}(t)\|^2_{\boldsymbol{L}^2(\Omega)}\right] + 2\nu \|\nabla \boldsymbol{u}(t)\|^2_{\boldsymbol{L}^2(\Omega)} = 0, \quad t \geq 0.$$

Using inequality (6), we obtain

$$\frac{d}{dt}\left[\|\boldsymbol{u}(t)\|^2_{\boldsymbol{L}^2(\Omega)} + \alpha^2 \|\nabla \boldsymbol{u}(t)\|^2_{\boldsymbol{L}^2(\Omega)}\right] + \frac{8\nu}{d_\Omega^2 + 4\alpha^2}\left[\|\boldsymbol{u}(t)\|^2_{\boldsymbol{L}^2(\Omega)} + \alpha^2 \|\nabla \boldsymbol{u}(t)\|^2_{\boldsymbol{L}^2(\Omega)}\right] \leq 0, \quad t \geq 0$$

and, hence,

$$\frac{d}{dt}\left[\exp\left(\frac{8\nu t}{d_\Omega^2 + 4\alpha^2}\right)\left\{\|\boldsymbol{u}(t)\|^2_{\boldsymbol{L}^2(\Omega)} + \alpha^2 \|\nabla \boldsymbol{u}(t)\|^2_{\boldsymbol{L}^2(\Omega)}\right\}\right] \leq 0, \quad t \geq 0. \tag{37}$$

Then, by integrating (37) with respect to $t$, we derive inequality (10). Thus, the proof of Theorem 1 is complete.

## 5. Concluding Remarks

In this paper, we prove the existence and uniqueness of a strong solution to the incompressible Navier–Stokes–Voigt model. The construction of a strong solution proceeds via the Faedo–Galerkin procedure with a special basis of eigenfunctions of the Stokes operator. Note that this approach allows easily obtaining approximations of strong solutions, which frequently reduce to approximate analytic or semi-analytic solutions when the flow domain has a simple symmetric shape. Such solutions favor a better understanding of the qualitative features of unsteady flows of viscoelastic fluids and can be used to test the relevant numerical, asymptotic, and approximate methods.

**Funding:** This research received no external funding.

**Conflicts of Interest:** The author declares no conflict of interest.

## Abbreviations

| Symbols and Notations | Meaning |
|---|---|
| $\Omega$ | flow domain |
| $\partial\Omega$ | boundary of $\Omega$ |
| $x_1, \ldots, x_n$ | space variables |
| $t$ | time |
| $\boldsymbol{u}$ | velocity field |
| $\boldsymbol{u}_0$ | initial velocity field |
| $p$ | pressure |
| $\nu$ | viscosity coefficient |
| $\alpha$ | relaxation coefficient |
| $\boldsymbol{f}$ | external forces field |
| $q$ | scalar potential for $\boldsymbol{f}$ |
| $T$ | fixed point in time |
| $\nabla$ | $\left(\frac{\partial}{\partial x_1}, \ldots, \frac{\partial}{\partial x_n}\right)$ |
| $\Delta$ | $\sum_{i=1}^{n} \frac{\partial^2}{\partial x_i^2}$ |
| $'$ | differentiation with respect to $t$ |
| $\oplus$ | orthogonal sum of subspaces |
| $\rightharpoonup$ | weak convergence |
| $\rightarrow$ | strong convergence |
| $\hookrightarrow$ | embedding |
| $A \times B$ | Cartesian product of two sets $A$ and $B$ |
| $\boldsymbol{x} \cdot \boldsymbol{y}$ | scalar product of vectors $\boldsymbol{x}, \boldsymbol{y} \in \mathbb{R}^n$ |
| $\boldsymbol{X} : \boldsymbol{Y}$ | scalar product of matrices $\boldsymbol{X}, \boldsymbol{Y} \in \mathbb{R}^{n \times n}$ |
| $(\boldsymbol{v}, \boldsymbol{w})_{\mathbf{H}}$ | scalar product of functions $\boldsymbol{v}$ and $\boldsymbol{w}$ from a Hilbert space $\mathbf{H}$ |
| $\\|\boldsymbol{v}\\|_{\mathbf{E}}$ | norm of function $\boldsymbol{v}$ from a Banach space $\mathbf{E}$ |
| $\mathcal{L}(E_1, E_2)$ | space of all bounded linear mappings from $E_1$ to $E_2$ |
| $\mathcal{D}(\Omega)$ | space of $\mathcal{C}^\infty$ functions with support contained in $\Omega$ |
| $\mathcal{V}(\Omega)$ | space of $\mathcal{C}^\infty$ divergence-free vector functions with support contained in $\Omega$ |
| $L^s(\Omega)$ | Lebesgue space |
| $H^k(\Omega)$ | Sobolev space |
| $\boldsymbol{V}^i(\Omega)$ | special Hilbert space defined by (2) for $i \in \{0, 1, 2\}$ |
| $\boldsymbol{G}(\Omega)$ | $\{\nabla h : \ h \in H^1(\Omega)\}$ |
| $H^1(\Omega)/\mathbb{R}$ | quotient of $H^1(\Omega)$ by $\mathbb{R}$ |
| $\sim$ | equivalence relation on $H^1(\Omega)$ |
| $\mathfrak{L}_d$ | layer with thickness $d$ |
| $d_\Omega$ | $\inf\{d : \ \Omega \subset \mathfrak{L}_d\}$ |
| $\bar{\xi}$ | $\{\omega \in H^1(\Omega) : \ \omega \sim \xi\}$ |
| $\mathbb{I}$ | embedding operator |
| $\mathbb{P}$ | Leray projection |
| $\mathbb{A}$ | Stokes operator |
| $\lambda_j$ | eigenvalue of Stokes operator |
| $\boldsymbol{w}_j$ | eigenfunction of Stokes operator |
| $\boldsymbol{v}_m$ | Galerkin solution |
| $C_i$ | positive constant independent of $m$ |

## References


1. Oskolkov, A.P. Solvability in the large of the first boundary value problem for a certain quasilinear third order system that is encountered in the study of the motion of a viscous fluid. *Zap. Nauchn. Semin. LOMI* **1972**, *27*, 145–160.
2. Oskolkov, A.P. On the uniqueness and solvability in the large of the boundary-value problems for the equations of motion of aqueous solutions of polymers. *Zap. Nauchn. Semin. LOMI* **1973**, *38*, 98–136.
3. Oskolkov, A.P. Some quasilinear systems that arise in the study of the motion of viscous fluids. *Zap. Nauchn. Semin. LOMI* **1975**, *52*, 128–157
4. Ladyzhenskaya, O.A. On some gaps in two of my papers on the Navier–Stokes equations and the way of closing them. *J. Math. Sci.* **2003**, *115*, 2789–2791. [CrossRef]
5. Sviridyuk, G.A. On a model of the dynamics of a weakly compressible viscoelastic fluid. *Russian Math. (Iz. VUZ)* **1994**, *38*, 59–68.
6. Sviridyuk, G.A.; Sukacheva T.G. On the solvability of a nonstationary problem describing the dynamics of an incompressible viscoelastic fluid. *Math. Notes* **1998**, *63*, 388–395. [CrossRef]
7. Korpusov, M.O.; Sveshnikov, A.G. Blow-up of Oskolkov's system of equations. *Sb. Math.* **2009**, *200*, 549–572. [CrossRef]
8. Ladyzhenskaya, O.A. On the global unique solvability of some two-dimensional problems for the water solutions of polymers. *J. Math. Sci.* **2000**, *99*, 888–897. [CrossRef]
9. Baranovskii, E.S. Mixed initial–boundary value problem for equations of motion of Kelvin–Voigt fluids. *Comput. Math. Math. Phys.* **2016**, *56*, 1363–1371. [CrossRef]
10. Baranovskii, E.S. Global solutions for a model of polymeric flows with wall slip. *Math. Meth. Appl. Sci.* **2017**, *40*, 5035–5043. [CrossRef]
11. Kaya, M.; Celebi, A.O. Existence of weak solutions of the g-Kelvin–Voight equation. *Math. Comput. Model.* **2009**, *49*, 497–504. [CrossRef]
12. Baranovskii, E.S. Flows of a polymer fluid in domain with impermeable boundaries. *Comput. Math. Math. Phys.* **2014**, *54*, 1589–1596. [CrossRef]
13. Berselli, L.C.; Spirito, S. Suitable weak solutions to the 3D Navier–Stokes equations are constructed with the Voigt approximation. *J. Differ. Equ.* **2017**, *262*, 3285–3316. [CrossRef]
14. Caffarelli, L.; Kohn, R.; Nirenberg, L. Partial regularity of suitable weak solutions of the Navier–Stokes equations. *Comm. Pure Appl. Math.* **1982**, *35*, 771–831. [CrossRef]
15. Fedorov, V.E.; Ivanova, N.D. Inverse problem for Oskolkov's system of equations. *Math. Meth. Appl. Sci.* **2017**, *40*, 6123–6126. [CrossRef]
16. Plekhanova, M.V.; Baybulatova, G.D.; Davydov, P.N. Numerical solution of an optimal control problem for Oskolkov's system. *Math. Meth. Appl. Sci.* **2018**, *41*, 9071–9080. [CrossRef]
17. Antontsev, S.N.; Khompysh, K. Kelvin–Voight equation with p-Laplacian and damping term: Existence, uniqueness and blow-up. *J. Math. Anal. Appl.* **2017**, *446*, 1255–1273. [CrossRef]
18. Artemov, M.A.; Baranovskii, E.S. Solvability of the Boussinesq approximation for water polymer solutions. *Mathematics* **2019**, *7*, 611. [CrossRef]
19. Mohan, M.T. On the three dimensional Kelvin–Voigt fluids: Global solvability, exponential stability and exact controllability of Galerkin approximations. *Evol. Equ. Control Theory* **2019**. [CrossRef]
20. Boyer, F.; Fabrie, P. *Mathematical Tools for the Study of the Incompressible Navier–Stokes Equations and Related Models*; Springer: New York, NY, USA, 2013. [CrossRef]
21. Temam, R. *Navier–Stokes Equations—Theory and Numerical Analysis*; North-Holland Publishing Co.: Amsterdam, The Netherlands, 1977.
22. Simon, J. Compact sets in the space $L^p(0,T;B)$. *Ann. Mat. Pura Appl.* **1986**, *146*, 65–96. [CrossRef]
23. Agranovich, M.S. *Sobolev Spaces, Their Generalizations, and Elliptic Problems in Smooth and Lipschitz Domains*; Springer: Cham, Switzerland, 2015. [CrossRef]

24. Galdi, G.P. *An Introduction to the Mathematical Theory of the Navier–Stokes Equations—Steady-State Problems*; Springer: New York, NY, USA, 2011. [CrossRef]
25. Siddiqi, A.H. *Functional Analysis and Applications*; Springer: Singapore, 2018. [CrossRef]

*Article*

# Nonlinear Spatiotemporal Viral Infection Model with CTL Immunity: Mathematical Analysis

**Jaouad Danane [1], Karam Allali [1], Léon Matar Tine [2,3,*] and Vitaly Volpert [2,3,4]**

[1] Laboratory of Mathematics and Applications, Faculty of Sciences and Technologies, University Hassan II of Casablanca, P.O.Box 146, Mohammedia, Morocco; jaouaddanane@gmail.com (J.D.); allali@hotmail.com (K.A.)

[2] CNRS UMR 5208 Institut Camille Jordan, University Claude Bernard Lyon 1, University de Lyon, 69622 Villeurbanne CEDEX, France; volpert@math.univ-lyon1.fr

[3] INRIA Team Dracula, INRIA Lyon La Doua, 69603 Villeurbanne, France

[4] S.M. Nikol'skii Mathematical Institute, Peoples' Friendship University of Russia (RUDN University), 6 Miklukho-Maklaya St, 117198 Moscow, Russia

* Correspondence: leon-matar.tine@univ-lyon1.fr

† Fully documented templates are available in the elsarticle package on CTAN.

Received: 27 November 2019; Accepted: 13 December 2019; Published: 1 January 2020

**Abstract:** A mathematical model describing viral dynamics in the presence of the latently infected cells and the cytotoxic T-lymphocytes cells (CTL), taking into consideration the spatial mobility of free viruses, is presented and studied. The model includes five nonlinear differential equations describing the interaction among the uninfected cells, the latently infected cells, the actively infected cells, the free viruses, and the cellular immune response. First, we establish the existence, positivity, and boundedness for the suggested diffusion model. Moreover, we prove the global stability of each steady state by constructing some suitable Lyapunov functionals. Finally, we validated our theoretical results by numerical simulations for each case.

**Keywords:** viral infection; diffusion; Lyapunov functional; convergence

## 1. Introduction

Viral infections represent a major cause of morbidity with important consequences for patient's health and for the society. Among the most dangerous, let us cite the human immunodeficiency virus (HIV) that attacks immune cells leading to the deficiency of the immune system [1,2], the human papillomavirus (HPV) that infects basal cells of the cervix [3,4], and the hepatitis B virus (HBV) and the hepatitis C virus (HCV) that attack liver cells [5–8]. Mathematical modeling becomes an important tool for the understanding and predicting the spread of viral infection, and for the development of efficient strategies to control its dynamics [9–12]. One of the basic models of viral infection suggested by Nowak in 1996 describes the interactions among uninfected cells, infected cells, and free viruses. Nowadays, modeling of viral infection actively develops with a variety of new models and methods [9–15] (see also the monograph [16] and the references therein). The action of immune system was introduced into the basic model with cytotoxic T-lymphocytes cells (CTL) killing infected cells [12–15,17,18]. The impact of CTL cells with a saturated incidence function was considered in [12]:

$$
\begin{cases}
\dot{H} = \lambda - d_1 H - \dfrac{k_1 H V}{H + V}, \\
\dot{S} = \dfrac{k_1 H V}{H + V} - d_2 S - k_2 S, \\
\dot{Y} = k_2 S - d_3 Y - p Y Z, \\
\dot{V} = a Y - d_4 V, \\
\dot{Z} = c Y Z - b Z.
\end{cases} \tag{1}
$$

Here $H$, $S$, $Y$, $V$, and $Z$ represent the densities of uninfected cells, exposed cells, infected cells, free virus, and CTL cells, respectively. Our model uses a more realistic saturated incidence function $\dfrac{k_1 H V}{H + V}$ [10,11]. This saturated incidence functional describes the infection rate taking into consideration the effect of free viruses crowd near the healthy cells. The parameters of the system in Equation (1) are described in Table 1.

**Table 1.** The parameters of the mathematical model and their descriptions.

| Coefficient | Description |
|---|---|
| $\lambda$ | The birth rate of the uninfected cells |
| $k_1$ | The rate of infection |
| $d_1$ | The natural mortality of the susceptible cells |
| $d_2$ | The death rate of exposed cells |
| $k_2$ | The average that exposed cells become infected |
| $d_3$ | The death rate of infected cells, not by CTL killing |
| $a$ | The rate of production the virus by infected cells |
| $d_4$ | The rate of viral clearance |
| $p$ | Clearance rate of infection |
| $c$ | Activation rate CTL cells |
| $b$ | Death rate of CTL cells |
| $d$ | Diffusion coefficient |

The majority of mathematical models of viral infection ignores the spatial movement of viruses and cells, assuming that the virus and cell populations are well mixed [19]. However, their mobility and a nonuniform spatial distribution can play an important role for the infection development [20]. Thus far, few studies have been devoted to the influence of spatial structure on the dynamics of the virus [21,22]. Reaction–diffusion waves of infection spreading were studied in [23–25].

In this work, we consider the previous model in Equation (1) taking into account virus diffusion:

$$
\begin{cases}
\dfrac{\partial H}{\partial t} = \lambda - d_1 H - \dfrac{k_1 H V}{H + V}, \\
\dfrac{\partial S}{\partial t} = \dfrac{k_1 H V}{H + V} - d_2 S - k_2 S, \\
\dfrac{\partial Y}{\partial t} = k_2 S - d_3 Y - p Y Z, \\
\dfrac{\partial V}{\partial t} = d \Delta V + a Y - d_4 V, \\
\dfrac{\partial Z}{\partial t} = c Y Z - b Z.
\end{cases} \tag{2}
$$

As in (1), $H, S, Y, V$, and $Z$ represent the densities of uninfected cells, latently infected cells, infected cells, free virus, and CTLs cells, depending now on the space coordinate location $x$ and on time $t$, $x \in \Omega$, where $\Omega$ is an open bounded set in $\mathbb{R}^n$. The positive constant $d$ is the virus diffusion coefficient. All parameters of the system in Equation (2) have the same biological meanings as in the model in Equation (1). In this paper, we consider the system in Equation (2) with the homogeneous Neumann boundary conditions:

$$\frac{\partial V}{\partial \nu} = 0, \qquad \text{on} \quad \partial\Omega \times [0, +\infty) \tag{3}$$

and the initial conditions:

$$H(x,0) = \varphi_1(x) \geq 0; S(x,0) = \varphi_2(x) \geq 0; Y(x,0) = \varphi_3(x) \geq 0; \tag{4}$$

$$V(x,0) = \varphi_4(x) \geq 0; Z(x,0) = \varphi_5(x) \geq 0 \quad \forall x \in \bar{\Omega}. \tag{5}$$

The paper is organized as follows. The next section is devoted to the well-posedness of the model, followed in Section 3 by the global stability analysis. In Section 4, we illustrate the results with the numerical simulations.

## 2. Well-Posedness of Model

In this section, we investigate the well-posedness of the model in Equation (2) proving the global existence, the positivity and the boundedness of solutions.

**Proposition 1.** *For any initial condition satisfying Equations (4)–(5), there exists a unique solution to the problem in Equations (2)–(3) defined on $t \in (0, +\infty)$. Moreover, this solution stays non-negative and bounded for all $t > 0$.*

**Proof.** Let consider the set

$$X_T = \{V \in L^2([0,T] \times \Omega) \cap C\left([0,T]; L^2(\Omega)\right); 0 \leq V(x,t) \leq C_T\},$$

where $T$ is a fixed positive constant and $C_T = \| V_0 \|_{L^\infty(\Omega)} + a\frac{\lambda}{m}T$ $(V_0 = V(x,0))$, where $m = min(d_1, d_2, d_3, pb/c)$.

To prove the existence of the solution, we define the following map

$$\begin{aligned} \Psi : X_T &\to L^2([0,T] \times \Omega) \\ \bar{V} &\mapsto \Psi(\bar{V}) = V \end{aligned}$$

such that

$$\begin{cases} \dfrac{\partial V}{\partial t} - d\Delta V = aY - d_4 V, & \text{in } \Omega \\ \dfrac{\partial V}{\partial \nu} = 0, & \text{in } [0,T] \times \partial\Omega \\ V(x,0) = \varphi_4(x), & \forall x \in \overline{\Omega}, \end{cases} \tag{6}$$

where $Y$ is the third component of the solution vector of the following subsystem

$$\begin{cases} \dfrac{\partial H(x,t)}{\partial t} = \lambda - d_1 H(x,t) - \dfrac{k_1 H(x,t)\bar{V}(x,t)}{H(x,t) + \bar{V}(x,t)}, \\ \dfrac{\partial S(x,t)}{\partial t} = \dfrac{k_1 H(x,t)\bar{V}(x,t)}{H(x,t) + \bar{V}(x,t)} - d_2 S(x,t) - k_2 S(x,t), \\ \dfrac{\partial Y(x,t)}{\partial t} = k_2 S(x,t) - d_3 Y(x,t) - pY(x,t)Z(x,t), \\ \dfrac{\partial Z(x,t)}{\partial t} = cY(x,t)Z(x,t) - bZ(x,t), \end{cases} \tag{7}$$

with the initial data

$$\begin{aligned}&H(x,0)=\varphi_1(x)\geq 0; S(x,0)=\varphi_2(x)\geq 0; Y(x,0)=\varphi_3(x)\geq 0;\\&Z(x,0)=\varphi_5(x)\geq 0 \quad \forall x\in\bar{\Omega}.\end{aligned}$$

Then, the system in Equation (7) can be written abstractly in $X=\Omega^4$ by the following form

$$\begin{aligned}&U'(t)=AU(t)+F(U(t)),\ \forall t>0,\\&U(0)=U_0\in X,\end{aligned}$$

with $U=(H,S,Y,Z)^T$, $U_0=(\varphi_1,\varphi_2,\varphi_3,\varphi_5)^T$, and

$$F(U(t))=\begin{pmatrix}\lambda-d_1H(x,t)-\dfrac{k_1H(x,t)\bar{V}(x,t)}{H(x,t)+\bar{V}(x,t)}\\ \dfrac{k_1H(x,t)\bar{V}(x,t)}{H(x,t)+\bar{V}(x,t)}-d_2S(x,t)-k_2S(x,t)\\ k_2S(x,t)-d_3Y(x,t)-pY(x,t)Z(x,t)\\ cY(x,t)Z(x,t)-bZ(x,t)\end{pmatrix}.$$

It is clear that $F$ is locally Lipschitz in $U$. Using the theorem of Cauchy–Lipschitz, we deduce that the system in Equation (7) admits a unique local solution on $[0,\tau]$, where $\tau\leq T$. In addition, the system in Equation (7) can be written of the form

$$\begin{cases}\dfrac{\partial H}{\partial t}=F_1(H,S,Y,Z),\\ \dfrac{\partial S}{\partial t}=F_2(H,S,Y,Z),\\ \dfrac{\partial Y}{\partial t}=F_3(H,S,Y,Z),\\ \dfrac{\partial Z}{\partial t}=F_4(H,S,Y,Z).\end{cases}$$

It is easy to see that the functions $F_i(H,S,Y,Z), 1\leq i\leq 4$ are continuously differentiable, verifying $F_1(0,S,Y,Z)=\lambda\geq 0$, $F_2(H,0,Y,Z)=\dfrac{k_1H\bar{V}}{H+\bar{V}}\geq 0$, $F_3(H,S,0,Z)=k_2S\geq 0$ and $F_4(H,S,Y,0)=0$ for all $H,S,Y,\bar{V},Z\geq 0$. Since the initial data of the system in Equation (7) are nonnegative, we obtain the positivity of $H$, $S$, $Y$, and $Z$ thanks to the quasi-reversibility principle.

Now, we show the boundedness of solution. Let

$$\begin{aligned}\mathcal{T}(x,t)&=H(x,t)+S(x,t)+Y(x,t)+\frac{p}{c}Z(x,t)\\ \frac{\partial\mathcal{T}}{\partial t}&=\lambda-d_1H(x,t)-d_2S(x,t)-d_3Y(x,t)-p\frac{b}{c}Z(x,t)\\ &\leq\lambda-m\mathcal{T}(x,t),\end{aligned}$$

Then,

$$\mathcal{T}(x,t)\leq\mathcal{T}(x,0)e^{-mt}+\frac{\lambda}{m}(1-e^{-mt}),$$

For biological reasons, we assume that the problem initial data are upper-bounded by the carrying capacity. This means that $\mathcal{T}(x,0)\leq\dfrac{\lambda}{m}$. We deduce that

$$H(x,t) \leq \frac{2\lambda}{m},\ S(x,t) \leq \frac{2\lambda}{m},$$
$$Y(x,t) \leq \frac{2\lambda}{m},\ Z(x,t) \leq \frac{2\lambda}{m}.$$

Thus, $H$, $S$, $Y$, and $Z$ are bounded.

Let us recast the system in Equation (6) as follows

$$\begin{cases} \dfrac{\partial V}{\partial t} - d\Delta V + d_4 V = aY, & \text{on } \Omega, \\ \dfrac{\partial V}{\partial \nu} = 0, & \text{on } \partial\Omega. \end{cases}$$

We know that $0 \leq Y(x,t) \leq \frac{\lambda}{m}$, then from the proposition 2.1 in [26], we deduce for all $V_0 \in L^2(\Omega)$ the existence and the uniqueness of the solution $V \in L^2([0,T];H^1(\Omega)) \cap C([0,T];L^2(\Omega))$ such that $\partial_t V \in L^2([0,T];H^1(\Omega)')$. Furthermore, if $V_0 \in L^\infty(\Omega)$, by using the maximum principle relation, we have

$$0 \leq V(t,x) \leq \| V_0 \|_{L^\infty(\Omega)} + a\frac{\lambda}{m}T.$$

We note that $\Psi$ is well defined and continuous and $\Psi(X_T)$ is compact, thus $\Psi$ admits a fixed point. Then, we conclude the existence of the solution $V$ of (6) and it is positive and bounded. □

*Equilibria and Basic Reproduction Number*

The system in Equation (2) has an infection-free equilibrium $E_f = (\frac{\lambda}{d_1}, 0, 0, 0, 0)$, corresponding to the total absence of viral infection. The basic reproduction number of the system in Equation (2) is given by

$$R_0 = k_1 \frac{k_2}{d_2 + k_2} \frac{a}{d_3} \frac{1}{d_4}, \tag{8}$$

with $\frac{k_2}{d_2 + k_2}$ the ratio of exposed cells that will become infected, $\frac{a}{d_3}$ the average of free virus produced by an infected cell, and $\frac{1}{d_4}$ the lifespan of the virus. The biological interpretation of $R_0$ represents the rate of secondary infections generated by an infected cell when it is introduced into a population of uninfected cells.

In addition to the disease free equilibrium, our system (Equation (2)) admits three endemic equilibria. The first of them is $E_1 = (H_1, S_1, Y_1, V_1, Z_1)$, where

$$H_1 = \frac{\lambda}{d_1 + k_1(1 - \frac{1}{R_0})},\ Z_1 = 0,$$
$$S_1 = \frac{k_1 \lambda R_0 (1 - \frac{1}{R_0})}{(d_2 + k_2)(d_1 + k_1(1 - \frac{1}{R_0}))((1 - \frac{1}{R_0})R_0 + 1)},$$
$$Y_1 = \frac{d_4 \lambda R_0 (1 - \frac{1}{R_0})}{ad_1 + ak_1(1 - \frac{1}{R_0})},\ V_1 = \frac{\lambda R_0 (1 - \frac{1}{R_0})}{d_1 + k_1(1 - \frac{1}{R_0})}.$$

This endemic steady state is specified as endemic equilibrium without cellular immunity. The second endemic steady state is $E_2 = (H_2, S_2, Y_2, V_2, Z_2)$, where

$$
\begin{aligned}
H_2 &= \frac{-abd_1 - abk_1 + \lambda c d_4 + \sqrt{A}}{2cd_1d_4}, \\
S_2 &= \frac{d_3 R_0}{k_2} \frac{b(-abd_1 - abk_1 + \lambda c d_4 + \sqrt{A})}{c(abd_1 - abk_1 + \lambda c d_4 + \sqrt{A})}, \\
Y_2 &= \frac{b}{c}, \; V_2 = \frac{ba}{cd_4}, \\
Z_2 &= \frac{d_3((R_0 - 1)(-ak_1 b + \lambda c d_4 + \sqrt{A}) - abd_1(R_0 + 1))}{p(abd_1 - ak_1 b + \lambda c d_4 + \sqrt{A})}.
\end{aligned}
$$

This endemic steady state is specified as endemic equilibrium with cellular immunity. It is also called interior equilibrium. The third endemic steady state is $E_3 = (H_3, S_3, Y_3, V_3, Z_3)$, where

$$
\begin{aligned}
H_3 &= -\frac{abd_1 + abk_1 - \lambda c d_4 + \sqrt{A}}{2cd_1d_4}, \\
S_3 &= \frac{k_1 H_3 V_3}{(k_2 + d_2)(H_3 + V_3)}, \\
Y_3 &= \frac{b}{c}, \; V_3 = \frac{ba}{cd_4}, \\
Z_3 &= \frac{k_2 S_3 - d_3 Y_3}{pY_3},
\end{aligned}
$$

with $A = (abk_1 - \lambda c d_4)^2 + a^2 b^2 d_1^2 + 2a^2 b^2 d_1 k_1 + 2\lambda abcd_1 d_4$.

Noting that $H_3 < 0$, this is not biologically relevant, thus the steady state $E_3$ is not considered. When $R_0 > 1$, the equilibrium $E_1$ exists. We define the reproduction rate of the CTL $R_{CTL}$ immune response by

$$
R_{CTL} = \frac{cY_1}{b} = \frac{cd_4 \lambda R_0 (1 - \frac{1}{R_0})}{abd_1 + abk_1 (1 - \frac{1}{R_0})}.
$$

Note that the endemic state $E_2$ exists when $R_{CTL} > 1$. Indeed, if one considers $R_0 > 1$ then, in total absence of CTL immune response, the infected cell loaded per unit time is $\frac{d_4 \lambda R_0 (1 - \frac{1}{R_0})}{ad_1 + ak_1 (1 - \frac{1}{R_0})}$. From the system in Equation (2), we have the CTL cells reproduced due to infected cells stimulating per unit time is $\frac{cd_4 \lambda R_0 (1 - \frac{1}{R_0})}{ad_1 + ak_1 (1 - \frac{1}{R_0})} = cY_1$. The CTL charge during the lifespan of a CTL cell is $\frac{cd_4 \lambda R_0 (1 - \frac{1}{R_0})}{abd_1 + abk_1 (1 - \frac{1}{R_0})} = R_{CTL}$. Thus, if $\frac{cd_4 \lambda R_0 (1 - \frac{1}{R_0})}{abd_1 + abk_1 (1 - \frac{1}{R_0})} > 1$, we deduce the existence of $E_2$.

## 3. Global Stability

To prove the global stability of the uninfected and the infected steady states, we use the method of construction of Lyapunov functions developed in [12] and can claim the following result

**Theorem 1.** *The disease-free equilibrium $E_f$ of the model in Equation (2) is globally asymptotically stable when $R_0 < 1$.*

**Proof.** We define the function $G_f$ by

$$
G_f(x, t) = S + \frac{d_2 + k_2}{k_2} Y + \frac{d_3(d_2 + k_2)}{ak_2} V + \frac{p}{c} \frac{d_2 + k_2}{k_2} Z.
$$

Then, by using the equations of the system in Equation (2), the time derivative of $G_f$ verifies

$$\frac{\partial G_f}{\partial t} \leq \frac{d_3 d_4(d_2+k_2)}{ak_2}(R_0-1)V.$$

Now, we define a Lyapunov function as follows

$$L_f = \int_\Omega G_f dx.$$

Calculating the time derivative of $L_f$ along the positive solutions of the model in Equation (2), we obtain

$$\begin{aligned}\frac{dL_f}{dt} &= \int_\Omega \frac{\partial G_f}{\partial t}dx \\ &\leq \int_\Omega \left(\frac{d_3 d_4(d_2+k_2)}{ak_2}(R_0-1)V\right)dx.\end{aligned}$$

Thus, if $R_0 < 1$ implies that $\frac{dL_f}{dt} \leq 0$. The largest compact invariant is

$$E = \{(H,S,Y,V,Z) \mid V = 0\},$$

according to LaSalle's invariance principle, $\lim_{t+\infty} V(x,t) = 0$, the limit system of equations is

$$\begin{cases}\frac{\partial H(x,t)}{\partial t} = \lambda - d_1 H(x,t), \\ \frac{\partial S(x,t)}{\partial t} = -d_2 S(x,t) - k_2 S(x,t), \\ \frac{\partial Y(x,t)}{\partial t} = k_2 S(x,t) - d_3 Y(x,t) - pY(x,t)Z(x,t), \\ \frac{\partial Z(x,t)}{\partial t} = cY(x,t)Z(x,t) - bZ(x,t).\end{cases}$$

For simplicity, we use the same notation,

$$G_f(H,S,Y,Z) = \frac{1}{H_0}\left(H - H_0 - H_0 \ln\frac{H}{H_0}\right) + S + \frac{d_2+k_2}{k_2}Y + \frac{p}{c}\frac{d_2+k_2}{k_2}Z.$$

Since $H_0 = \frac{\lambda}{d_1}$,

$$\frac{\partial G_f}{\partial t}(H,S,Y,Z) = d_1\left(2 - \frac{H}{H_0} - \frac{H_0}{H}\right) - \frac{d_3(d_2+k_2)}{k_2}Y - \frac{pb}{c}\frac{d_2+k_2}{k_2}Z.$$

We define another Lyapunov function

$$L_f = \int_\Omega G_f dx,$$

then, the time derivative of $L_f$ satisfies

$$\begin{aligned}\frac{dL_f}{dt} &= \int_\Omega \frac{\partial G_f}{\partial t}dx \\ &\leq \int_\Omega \left(d_1\left(2 - \frac{H}{H_0} - \frac{H_0}{H}\right) - \frac{d_3(d_2+k_2)}{k_2}Y - \frac{pb}{c}\frac{d_2+k_2}{k_2}Z\right)dx.\end{aligned}$$

Since the arithmetic mean is greater than or equal to the geometric mean, it follows

$$2-\frac{H}{H_0}-\frac{H_0}{H}\leq 0,$$

therefore $\frac{dL_f}{dt}\leq 0$ and the equality holds if $H=H_0$ and $S=Y=Z=0$, which completes the proof. □

Now, we are interested in the stability of the infected steady state $E_1$. Let us state the following theorem.

**Theorem 2.** *The infected steady state $E_1$ of the model in Equation (2) is globally asymptotically stable when $R_{CTL}\leq 1<R_0$. In this case, the other infected steady state $E_2$ does not exist.*

**Proof.** Firstly, we define the function

$$\begin{aligned}G_1(x,t)&=H-H_1-\int_{H_1}^{H}\frac{(d_2+k_2)S_1}{\frac{k_1uV_1}{u+V_1}}du+S-S_1-S_1\ln\frac{S}{S_1}\\&+\frac{d_2+k_2}{k_2}\left(Y-Y_1-Y_1\ln\left(\frac{Y}{Y_1}\right)\right)+\frac{d_3(d_2+k_2)}{ak_2}\left(V-V_1-V_1\ln\left(\frac{V}{V_1}\right)\right)\\&+\frac{p}{c}\frac{d_2+k_2}{k_2}Z.\end{aligned}$$

Using the same technique proposed in [12], we obtain

$$\begin{aligned}\frac{\partial G_1}{\partial t}&=-\frac{d_1H_1}{H(H_1+V_1)}(H-H_1)^2\\&-(d_2+k_2)S_1\left(\frac{H(V-V_1)^2}{V_1(H+V_1)(H+V)}\right)\\&+(d_2+k_2)S_1\left(5-\frac{H_1}{H}\frac{H+V_1}{H_1+V_1}-\frac{S_1}{S}\frac{HV}{H_1V_1}\frac{H_1+V_1}{H+V}-\frac{SY_1}{S_1Y}-\frac{YV_1}{Y_1V}-\frac{H+V}{H+V_1}\right)\\&+pZ\frac{d_2+k_2}{k_2}\frac{b}{c}(R_{CTL}-1).\end{aligned}$$

Now, let us consider the following Lyapunov function

$$L_1=\int_\Omega G_1dx,$$

then we deduce

$$\begin{aligned}\frac{dL_1}{dt}&=\int_\Omega\frac{\partial G_1}{\partial t}dx\\&=\int_\Omega\left(-\frac{d_1H_1}{H(H_1+V_1)}(H-H_1)^2-(d_2+k_2)S_1\left(\frac{H(V-V_1)^2}{V_1(H+V_1)(H+V)}\right)\right.\\&+(d_2+k_2)S_1\left(5-\frac{H_1}{H}\frac{H+V_1}{H_1+V_1}-\frac{S_1}{S}\frac{HV}{H_1V_1}\frac{H_1+V_1}{H+V}-\frac{SY_1}{S_1Y}-\frac{YV_1}{Y_1V}-\frac{H+V}{H+V_1}\right)\\&\left.+pZ\frac{d_2+k_2}{k_2}\frac{b}{c}(R_{CTL}-1)\right)dx.\end{aligned}$$

Again, since the arithmetic mean is greater than or equal to the geometric mean, it follows

$$5-\frac{H_1}{H}\frac{H+V_1}{H_1+V_1}-\frac{S_1}{S}\frac{HV}{H_1V_1}\frac{H_1+V_1}{H+V}-\frac{SY_1}{S_1Y}-\frac{YV_1}{Y_1V}-\frac{H+V}{H+V_1}\leq 0.$$

In addition, when $R_{CTL}<1$, which means that $\frac{dL_1}{dt}\leq 0$.

Therefore, by Lyapunov–LaSalle invariance theorem, $E_1$ is globally asymptotically stable when $R_0>1$ and $R_{CTL}\leq 1$. □

To prove the stability of $E_2$ equilibrium, let us state the following theorem

**Theorem 3.** *The infected steady state $E_2$ of the model in Equation (2) is globally asymptotically stable when $R_0>1$ and $R_{CTL}>1$. In this case, the other infected steady state $E_1$ is unstable.*

**Proof.** We consider the following function

$$\begin{aligned}G_2(x,t)&=H-H_2-\int_{H_2}^{H}\frac{(d_2+k_2)S_2}{\frac{k_1uV_2}{u+V_2}}du+S-S_2-S_2\ln\frac{S}{S_2}\\&+\frac{d_2+k_2}{k_2}\left(Y-Y_2-Y_2\ln\left(\frac{Y}{Y_2}\right)\right)+\frac{d_3(d_2+k_2)+(d_2+k_2)pZ_2}{ak_2}\\&\times\left(V-V_2-V_2\ln\left(\frac{V}{V_2}\right)\right)+\frac{p}{c}\frac{d_2+k_2}{k_2}\left(Z-Z_2-Z_2\ln\left(\frac{Z}{Z_2}\right)\right).\end{aligned}$$

Then, we have

$$\begin{aligned}\frac{\partial G_2}{\partial t}&=-\frac{d_1V_2}{H(H_2+V_2)}(H-H_2)^2\\&-(d_2+k_2)S_2\left(\frac{H(V-V_2)^2}{V_2(H+V_2)(H+V)}\right)\\&+(d_2+k_2)S_2\left(5-\frac{H_2}{H}\frac{H+V_2}{H_2+V_2}-\frac{S_2}{S}\frac{HV}{H_2V_2}\frac{H_2+V_2}{H+V}-\frac{SY_2}{S_2Y}-\frac{YV_2}{Y_2V}-\frac{H+V}{H+V_2}\right).\end{aligned}$$

As a result, we define a Lyapunov function as follows

$$L_2=\int_{\Omega}G_2dx,$$

then,

$$\begin{aligned}\frac{dL_2}{dt}&=\int_{\Omega}\frac{\partial G_2}{\partial t}dx\\&=\int_{\Omega}\left(-\frac{d_1V_2}{H(H_2+V_2)}(H-H_2)^2-(d_2+k_2)S_2\left(\frac{H(V-V_2)^2}{V_2(H+V_2)(H+V)}\right)\right.\\&\left.+(d_2+k_2)S_2\left(5-\frac{H_2}{H}\frac{H+V_2}{H_2+V_2}-\frac{S_2}{S}\frac{HV}{H_2V_2}\frac{H_2+V_2}{H+V}-\frac{SY_2}{S_2Y}-\frac{YV_2}{Y_2V}-\frac{H+V}{H+V_2}\right)\right)dx,\end{aligned}$$

since the arithmetic mean is greater than or equal to the geometric mean, it follows

$$5-\frac{H_2}{H}\frac{H+V_2}{H_2+V_2}-\frac{S_2}{S}\frac{HV}{H_2V_2}\frac{H_2+V_2}{H+V}-\frac{SY_2}{S_2Y}-\frac{YV_2}{Y_2V}-\frac{H+V}{H+V_2}\leq 0,$$

which means that $\frac{dL_2}{dt}\leq 0$, and the equality hods when $H=H_2$, $S=S_2$, $Y=Y_2$, $V=V_2$, and $Z=Z_2$. By the LaSalle invariance principle, the endemic point $E_2$ is globally stable. □

## 4. Numerical Simulations

In this section, we present the results of numerical simulations to validate the theoretical results of the previous section. We used the finite difference numerical method with Euler explicit scheme. The convergence of our numerical method was tested by successively decreasing the time and space steps. The values of parameters are given in Appendix A.

We considered the one-dimensional interval $0 \le x \le L$ and time $0 < t \le T$, where $L = 20$ (dimensionless space units) and $T = 200$ days. The initial conditions were chosen space dependent to illustrate behavior of spatially inhomogeneous solutions. We used an explicit numerical method with the space step $h_x = 0.01$ and time step $h_t = 0.1$. The program was implemented with Matlab (2014a, MathWorks, Natick, MA, USA).

Figure 1 shows spatiotemporal dynamics of uninfected cells (left) and the maximal and the minimal values of the virus concentration in space as a function of time (right). For the values of parameters considered in this example, the basic reproduction number $R_0 = 0.22 < 1$, which implies that the virus-free equilibrium is stable. Therefore, as expected, solution converges toward the equilibrium $E_f = (8.27 \times 10^2, 0, 0, 0, 0)$.

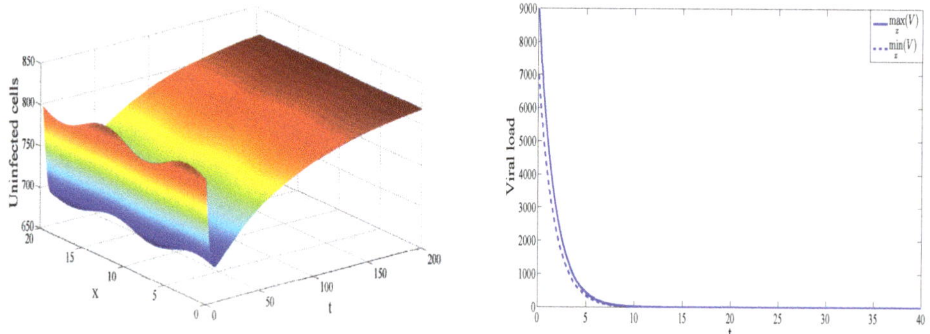


**Figure 1.** Dynamics of solution for $\lambda = 10$, $d_1 = 0.0139$, $k_1 = 0.04$, $d_2 = 0.0495$, $k_2 = 1.1$, $d_3 = 0.5776$, $a = 2$, $d_4 = 0.6$, $p = 0.0024$, $c = 0.15$, and $b = 0.5$. The concentration of uninfected cells is shown as a function of $x$ and $t$ (**left**). The maximal and the minimal value of the virus concentration with respect to $x$ are shown as functions of time (**right**).

To illustrate convergence to the endemic equilibrium point $E_1$, we considered the values of parameters presented in Figure 2. In this case, the basic reproduction number is greater than 1, $R_0 = 11.05 > 1$, and the immune reproduction number is less than 1, $R_{CTL} = 3.596 \times 10^{-1} < 1$. Therefore, the free-immune endemic equilibrium $E_1 = (19.96, 5.98 \times 10^{-1}, 1.14, 199.78, 0)$ is globally stable.

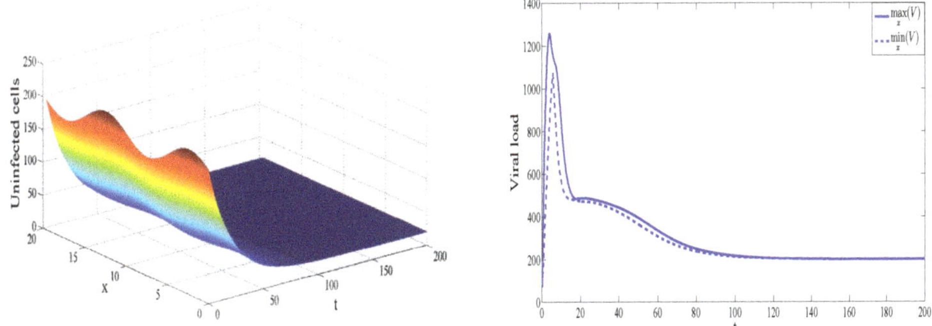


**Figure 2.** Dynamics of solution for $\lambda = 1$, $d_1 = 0.0139$, $k_1 = 0.04$, $d_2 = 0.0495$, $k_2 = 1.1$, $d_3 = 0.5776$, $a = 100$, $d_4 = 0.6$, $p = 0.0024$, $c = 0.15$, and $b = 0.5$. The concentration of uninfected cells is shown as a function of $x$ and $t$ (**left**). The maximal and the minimal value of the virus concentration with respect to $x$ are shown as functions of time (**right**).

In the case considered in Figure 3, we obtain $R_0 = 11.05 > 1$ and $R_{CTL} = 4.13 > 1$. Consequently, we observe that the endemic equilibrium $E_2 = (285.12, 6.55, 3.33, 555.55, 660.86)$ is stable. This, numerical results support the theoretical findings.

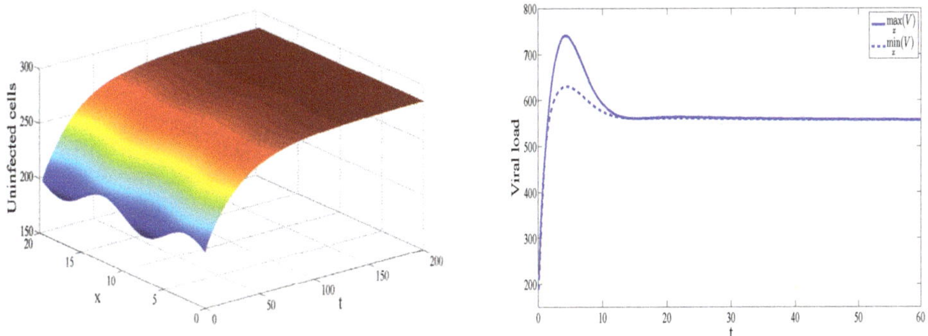


**Figure 3.** Dynamics of solution for $\lambda = 10$, $d_1 = 0.0139$, $k_1 = 0.04$, $d_2 = 0.0495$, $k_2 = 1.1$, $d_3 = 0.5776$, $a = 100$, $d_4 = 0.6$, $p = 0.0024$, $c = 0.15$, and $b = 0.5$. The concentration of uninfected cells is shown as a function of $x$ and $t$ (**left**). The maximal and the minimal value of the virus concentration with respect to $x$ are shown as functions of time (**right**).

## 5. Discussion and Conclusions

In this work, we study a model of viral infection in the presence of CTL cells and latently infected cells. We take into consideration not only the variation in time, but also spatial variation of virus distribution and its diffusion, where uninfected cells, latently infected cells, infected cells, and CTL cells do not exhibit any such spatial mobility. As a result, we show the stability of the free-disease equilibrium using the construction of a Lyapunov function when the reproduction number is less than one $(R_0 < 1)$. If $R_0 > 1$, then two scenarios are established. If $R_{CTL} < 1$, then the equilibrium point $E_1$ is globally asymptotically stable, while, for $R_{CTL} > 1$, the equilibrium point $E_2$ is globally asymptotically stable. We also performed numerical simulations of our model to illustrate behavior of solutions and to confirm the theoretical results. It was established that spatial diffusion of free viruses has no effect on the stability of the steady states. However, the effect appears only for the first days of infection observation.

**Author Contributions:** All authors contributed equally to this work. All authors have read and agreed to the published version of the manuscript.

**Funding:** This research was funded by Centre National de la Recherche Scientifique, grant number PICS 244832.

**Acknowledgments:** The work was partially supported by the France–Morocco program PICS 244832, "RUDN University Program 5-100" and the French–Russian project PRC2307.

**Conflicts of Interest:** The authors declare no conflict of interest.

## Appendix A. The Values of Parameters Used in the Numerical Simulations

The values of parameters used in the numerical simulations are given in Table A1:

**Table A1.** The values, units and meaning of all used parameters.

| Parameters | Units | Meaning | Value | References |
|---|---|---|---|---|
| $\lambda$ | cells $\mu L^{-1}$ day$^{-1}$ | Source rate of CD4+ T cells | $[0, 10]$ | [27] |
| $k_1$ | $\mu L$ virion$^{-1}$ day$^{-1}$ | Average of infection | $[2.5 \times 10^{-4}, 0.5]$ | [12] |
| $d_1$ | day$^{-1}$ | Decay rate of healthy cells | 0.0139 | [12] |
| $d_2$ | day$^{-1}$ | Death rate of exposed CD4+ T cells | 0.0495 | [12] |
| $k_2$ | day$^{-1}$ | The rate that exposed cells become infected CD4+ T cells | 1.1 | [12] |
| $d_3$ | day$^{-1}$ | Death rate of infected CD4+ T cells, not by CTL killing | 0.5776 | [12] |
| $a$ | day$^{-1}$ | The rate of production the virus by infected CD4+ T cells | $[2, 1250]$ | [12] |
| $d_4$ | day$^{-1}$ | Clearance rate of virus | $[0.3466, 2.4]$ | [12] |
| $p$ | $\mu L$ cell$^{-1}$ day$^{-1}$ | Clearance rate of infection | 0.0024 | [28] |
| $c$ | cells cell$^{-1}$ day$^{-1}$ | Activation rate CTL cells | 0.15 | [28] |
| $b$ | day$^{-1}$ | Death rate of CTL cells | 0.5 | [28] |
| $d$ | mm$^2$day$^{-1}$ | Diffusion coefficient | 0.01 | – |

## References

1. Perelson, A.S.; Neumann, A.U.; Markowitz, M.; Leonard, J.M.; Ho, D.D. HIV-1 dynamics in vivo: Virion clearance rate, infected cell life-span, and viral generation time. *Science* **1996**, *271*, 1582–1586. [CrossRef] [PubMed]
2. Adams, B.M.; Banks, H.T.; Davidian, M.; Kwon, H.D.; Tran, H.T.; Wynne, S.N.; Rosenberg, E.S. HIV dynamics: Modeling, data analysis, and optimal treatment protocols. *J. Comput. Appl. Math.* **2005**, *184*, 10–49. [CrossRef]
3. Burchell, A.N.; Winer, R.L.; de Sanjosé, S.; Franco, E.L. Epidemiology and transmission dynamics of genital HPV infection. *Vaccine* **2006**, *24*, S52–S61. [CrossRef] [PubMed]
4. Elbasha, E.H.; Dasbach, E.J.; Insinga, R.P. A multi-type HPV transmission model. *Bull. Math. Biol.* **2008**, *70*, 2126–2176. [CrossRef] [PubMed]
5. Li, M.; Zu, J. The review of differential equation models of HBV infection dynamics. *J. Virol. Methods* **2019**, *266*, 103–113. [CrossRef] [PubMed]
6. Meskaf, A.; Allali, K.; Tabit, Y. Optimal control of a delayed hepatitis B viral infection model with cytotoxic T-lymphocyte and antibody responses. *Int. J. Dyn. Control.* **2017**, *5*, 893–902. [CrossRef]
7. Wodarz, D. Hepatitis C virus dynamics and pathology: The role of CTL and antibody responses. *J. Gen. Virol.* **2003**, *84*, 1743–1750. [CrossRef]
8. Layden, J.E.; Layden, T.J. How can mathematics help us understand HCV? *Gastroenterology* **2001**, *120*, 1546–1549. [CrossRef]
9. Nowak, M.A.; Bangham, C.R.M. Population dynamics of immune responses to persistent viruses. *Science* **1996**, *272*, 74–79. [CrossRef]
10. Sun, Q.; Min, L.; Kuang, Y. Global stability of infection-free state and endemic infection state of a modified human immunodeficiency virus infection model. *IET Syst. Biol.* **2015**, *9*, 95–103. [CrossRef]
11. Sun, Q.; Min, L. Dynamics Analysis and Simulation of a Modified HIV Infection Model with a Saturated Infection Rate. *Comput. Math. Methods Med.* **2014**. [CrossRef] [PubMed]
12. Allali, K.; Danane, J.; Kuang, Y. Global Analysis for an HIV Infection Model with CTL Immune Response and Infected Cells in Eclipse Phase. *Appl. Sci.* **2017**, *7*, 861. [CrossRef]

13. Smith, H.L.; De Leenheer, P. Virus dynamics: A global analysis. *SIAM J. Appl. Math.* **2003**, *63*, 1313–1327. [CrossRef]
14. Daar, E.S.; Moudgil, T.; Meyer, R.D.; Ho, D.D. Transient highlevels of viremia in patients with primary human immunodeficiency virus type 1. *N. Engl. J. Med.* **1991**, *324*, 961–964. [CrossRef]
15. Kahn, J.O.; Walker, B.D. Acute human immunodeficiency virus type 1 infection. *N. Engl. J. Med.* **1998**, *339*, 33–39. [CrossRef]
16. Bocharov, G.; Volpert, V.; Ludewig, B.; Meyerhans, A. *Mathematical Immunology of Virus Infections*; Springer International Publishing: Cham, Switzerland, 2018; doi:10.1007/978-3-319-72317-4; [CrossRef]
17. Kaufmann, G.R.; Cunningham, P.; Kelleher, A.D.; Zauders, J.; Carr, A.; Vizzard, J.; Law, M.; Cooper, D.A. The Sydney Primary HIV Infection Study Group. Patterns of viral dynamics during primary human immunodeficiency virus type 1 infection. *J. Infect. Dis.* **1998**, *178*, 1812–1815. [CrossRef]
18. Schacker, T.; Collier, A.; Hughes, J.; Shea, T.; Corey, L. Clinical and epidemiologic features of primary HIV infection. *Ann. Int. Med.* **1996**, *125*, 257–264. [CrossRef]
19. Britton, N.F. *Essential Mathematical Biology*; Springer: London, UK, 2003.
20. Grebennikov, D.; Bouchnita, A.; Volpert, V.; Bessonov, N.; Meyerhans, A.; Bocharov, G. Spatial lymphocyte dynamics in lymph nodes predicts the CTL frequency needed for HIV infection control. *Front. Immunol.* **2019**, *10*, 1213. [CrossRef]
21. Maziane, M.; Hattaf, K.; Yousfi, N. Dynamics of a class of HIV infection models with cure of infected cells in eclipse stage. *Acta Biotheor.* **2015**, *63*, 363–380. [CrossRef]
22. Hattaf, K.; Yousfi, N. A numerical method for delayed partial differential equations describing infectious diseases. *Comput. Math. Appl.* **2016**, *72*, 2741–2750. [CrossRef]
23. Trofimchuk, S.; Volpert, V. Traveling waves for a bistable reaction-diffusion equation with delay. *SIAM J. Math. Anal.* **2018**, *50*, 1175–1199. [CrossRef]
24. Bocharov, G.; Meyerhans, A.; Bessonov, N.; Trofimchuk, S.; Volpert, V. Interplay between reaction and diffusion processes in governing the dynamics of virus infections. *J. Theor. Biol.* **2018**, *457*, 221–236. [CrossRef] [PubMed]
25. Bessonov, N.; Bocharov, G.; Touaoula, T.M.; Trofimchuk, S.; Volpert, V. Delay reaction-diffusion equation for infection dynamics. *Discret. Contin. Dyn. Syst.-B* **2019**, *24*, 2073–2091. [CrossRef]
26. Goudon, T.; Lagoutiere, F.; Tine, L.M. The Lifschitz-Slyozov equation with space-diffusion of monomers. *Kinet. Relat. Model.* **2012**, *5*, 325–355. [CrossRef]
27. Wang, Y.; Zhou, Y.; Wu, J.; Heffernan, J. Oscillatory viral dynamics in a delayed HIV pathogenesis model. *Math. Biosci.* **2009**, *219*, 104–112. [CrossRef]
28. Zhu, H.; Luo, Y.; Chen, M. Stability and Hopf bifurcation of a HIV infection model with CTL-response delay. *Comput. Math. Appl.* **2011**, *62*, 3091–3102. [CrossRef]

*Article*

# Nonlinear Multigrid Implementation for the Two-Dimensional Cahn–Hilliard Equation

**Chaeyoung Lee [1], Darae Jeong [2], Junxiang Yang [1] and Junseok Kim [1,*]**

1 Department of Mathematics, Korea University, Seoul 02841, Korea; chae1228@korea.ac.kr (C.L.); nexusxiang@outlook.com (J.Y.)
2 Department of Mathematics, Kangwon National University, Chuncheon-si 200-090, Korea; tinayoyo@kangwon.ac.kr
* Correspondence: cfdkim@korea.ac.kr

Received: 19 December 2019; Accepted: 4 January 2020; Published: 7 January 2020

**Abstract:** We present a nonlinear multigrid implementation for the two-dimensional Cahn–Hilliard (CH) equation and conduct detailed numerical tests to explore the performance of the multigrid method for the CH equation. The CH equation was originally developed by Cahn and Hilliard to model phase separation phenomena. The CH equation has been used to model many interface-related problems, such as the spinodal decomposition of a binary alloy mixture, inpainting of binary images, microphase separation of diblock copolymers, microstructures with elastic inhomogeneity, two-phase binary fluids, in silico tumor growth simulation and structural topology optimization. The CH equation is discretized by using Eyre's unconditionally gradient stable scheme. The system of discrete equations is solved using an iterative method such as a nonlinear multigrid approach, which is one of the most efficient iterative methods for solving partial differential equations. Characteristic numerical experiments are conducted to demonstrate the efficiency and accuracy of the multigrid method for the CH equation. In the Appendix, we provide C code for implementing the nonlinear multigrid method for the two-dimensional CH equation.

**Keywords:** Cahn–Hilliard equation; multigrid method; unconditionally gradient stable scheme

**MSC:** 65N06; 65N55

---

## 1. Introduction

In this paper, we consider a detailed multigrid [1] implementation of the following two-dimensional Cahn–Hilliard (CH) equation [2] and provide its C source code:

$$\begin{aligned} \frac{\partial \phi(x,y,t)}{\partial t} &= M\Delta\mu(x,y,t), \quad (x,y)\in\Omega, \quad t>0, \\ \mu(x,y,t) &= F'(\phi(x,y,t)) - \epsilon^2\Delta\phi(x,y,t), \end{aligned}$$

where $\phi$ is a conserved scalar field; $M$ is the mobility; $F(\phi) = 0.25(\phi^2-1)^2$ is the free energy function (see Figure 1); $\epsilon$ is the gradient interfacial energy coefficient; and $\Omega \subset \mathbb{R}^2$ is a bounded domain.

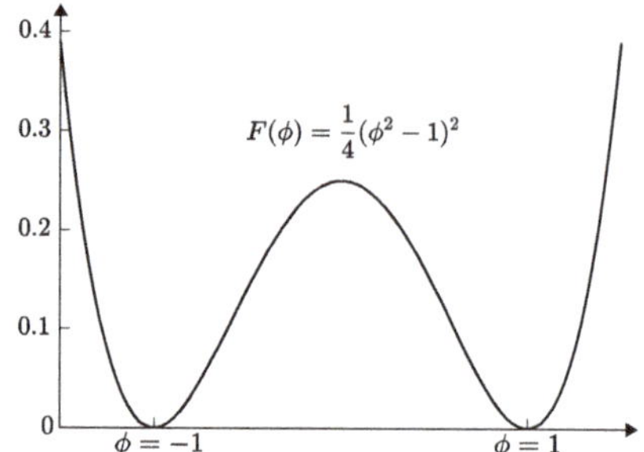


**Figure 1.** Double-well potential $F(\phi) = 0.25(\phi^2 - 1)^2$.

The homogeneous Neumann boundary conditions are used and are given as follows:

$$\mathbf{n} \cdot \nabla \phi = 0, \tag{1}$$
$$\mathbf{n} \cdot \nabla \mu = 0 \text{ on } \partial\Omega. \tag{2}$$

Here, $\mathbf{n}$ is the unit normal vector on the domain boundary $\partial\Omega$. The first boundary condition (1) implies that the interface contacts the domain boundary at a 90° angle. The second boundary condition (2) implies that the total mass is conserved.

We can derive the CH equation from the following total free energy functional

$$\mathcal{E}(\phi) = \int_\Omega \left[ F(\phi) + \frac{\epsilon^2}{2} |\nabla \phi|^2 \right] d\mathbf{x}.$$

Taking the variational derivative of $\mathcal{E}$ with respect to $\phi$, we define the chemical potential:

$$\mu = \frac{\delta \mathcal{E}}{\delta \phi} = F'(\phi) - \epsilon^2 \Delta \phi.$$

Conservation of mass implies the following CH equation

$$\phi_t = -\nabla \cdot \mathcal{F},$$

where the flux is given by $\mathcal{F} = -M\nabla\mu$. If we differentiate $\mathcal{E}(\phi)$ and $\int_\Omega \phi \, d\mathbf{x}$ with respect to time $t$, then we have

$$\begin{aligned} \frac{d}{dt}\mathcal{E}(\phi) &= \int_\Omega \left[ F'(\phi)\phi_t + \epsilon^2 \nabla\phi \cdot \nabla\phi_t \right] \mathrm{d}\mathbf{x} = \int_\Omega \mu\phi_t \, \mathrm{d}\mathbf{x} = M \int_\Omega \mu\Delta\mu \, \mathrm{d}\mathbf{x} \\ &= M \int_{\partial\Omega} \mu\nabla\mu \cdot \mathbf{n} \, \mathrm{d}s - M \int_\Omega \nabla\mu \cdot \nabla\mu \, \mathrm{d}\mathbf{x} = -M \int_\Omega |\nabla\mu|^2 \, \mathrm{d}\mathbf{x} \le 0 \end{aligned} \tag{3}$$

and

$$\frac{d}{dt} \int_\Omega \phi \, \mathrm{d}\mathbf{x} = \int_\Omega \phi_t \, \mathrm{d}\mathbf{x} = M \int_\Omega \Delta\mu \, \mathrm{d}\mathbf{x} = M \int_{\partial\Omega} \nabla\mu \cdot \mathbf{n} \, \mathrm{d}s = 0, \tag{4}$$

which imply that the total energy is decreasing and that the total mass is conserved in time, respectively. The CH equation was originally developed by Cahn and Hilliard to model spinodal decomposition in a binary alloy. The CH equation has been used to address many major problems such as the spinodal decomposition of a binary alloy mixture [3,4], inpainting of binary images [5,6], microphase separation of diblock copolymers [7,8], microstructures with elastic inhomogeneity [9,10], two-phase binary fluids [11,12], tumor growth models [13–15] and structural topology optimization [14,16]. Further details regarding the basic principles and practical applications of the CH equation are available in a recent review [14]. Thus, knowing how to implement a discrete scheme for the CH equation in detail is very useful because this equation is a building-block equation for many applications. The CH equation is discretized by using Eyre's unconditionally gradient stable scheme [17] and is solved by using a nonlinear multigrid technique [1], which is one of the most efficient iterative methods for solving partial differential equations. Several studies have used the nonlinear multigrid method for the CH-type equations [18–23]. However, details regarding the implementation, multigrid performance, and source codes have not been provided.

Therefore, the main purpose of this paper is to describe a detailed multigrid implementation of the two-dimensional CH equation, evaluate its performance and provide its C programming language source code.

The remainder of this paper is organized as follows. In Section 2, we describe the numerical solution in detail. In Section 3, we describe the characteristic numerical experiments that are conducted to demonstrate the accuracy and efficiency of the multigrid method for the CH equation. In Section 4, we provide a conclusion. In the Appendix A, we provide the C code for implementing the nonlinear multigrid technique for the two-dimensional CH equation.

## 2. Numerical Solution

We consider a finite difference approximation for the CH equation. An unconditionally gradient energy stable scheme, which was proposed by Eyre, is applied to the time discretization. A nonlinear multigrid technique [1] is applied to solve the resulting system at an implicit time level.

### 2.1. Discretization

We discretize the CH equation in the two-dimensional space $\Omega = (a,b) \times (c,d)$. Let $N_x = 2^p$ and $N_y = 2^q$ be the numbers of mesh points with integers $p$ and $q$. Let $\Delta x = (b-a)/N_x$ and $\Delta y = (d-c)/N_y$ be the mesh size. Let $\Omega_{ij} = \{(x_i, y_j) : x_i = a + (i-0.5)\Delta x,\ y_j = c + (j-0.5)\Delta y,\ 1 \le i \le N_x,\ 1 \le j \le N_y\}$ be a discrete computational domain. Let $\phi_{ij}^n$ and $\mu_{ij}^n$ be approximations of $\phi(x_i, y_j, t_n)$ and $\mu(x_i, y_j, t_n)$, respectively. Here, $t_n = n\Delta t$ and $\Delta t$ represent the temporal step. We assume a uniform mesh grid $h = \Delta x = \Delta y$ and a constant mobility $M = 1$. Using the nonlinear stabilized splitting scheme of Eyre's unconditionally gradient stable scheme, the CH equation is discretized as

$$\frac{\phi_{ij}^{n+1} - \phi_{ij}^n}{\Delta t} = \Delta_h \mu_{ij}^{n+1}, \tag{5}$$

$$\mu_{ij}^{n+1} = (\phi_{ij}^{n+1})^3 - \phi_{ij}^n - \epsilon^2 \Delta_h \phi_{ij}^{n+1}, \tag{6}$$

where the discrete Laplace operator is defined by $\Delta_h \psi_{ij} = (\psi_{i+1,j} + \psi_{i-1,j} + \psi_{i,j+1} + \psi_{i,j-1} - 4\psi_{ij})/h^2$. The homogeneous Neumann boundary conditions (1) and (2) are discretized as

$$\phi_{0j} = \phi_{1j},\ \phi_{N_x+1,j} = \phi_{N_x,j},\ \mu_{0j} = \mu_{1j},\ \mu_{N_x+1,j} = \mu_{N_x,j},\ j = 1, \ldots, N_y, \tag{7}$$

$$\phi_{i0} = \phi_{i1},\ \phi_{i,N_y+1} = \phi_{i,N_y},\ \mu_{i0} = \mu_{i1},\ \mu_{i,N_y+1} = \mu_{i,N_y},\ i = 1, \ldots, N_x. \tag{8}$$

We define the discrete residual as

$$r_{ij} = \Delta_h \mu_{ij}^{n+1} - \frac{\phi_{ij}^{n+1} - \phi_{ij}^n}{\Delta t}. \tag{9}$$

For each element $a_{ij}$ of size $N_x \times N_y$ in matrix A, we define the Frobenius norm with a scaling and infinite norm as

$$\|A\|_F = \sqrt{\frac{\sum_{i=1}^{N_x} \sum_{j=1}^{N_y} |a_{ij}|^2}{N_x N_y}} \text{ and } \|A\|_\infty = \max_{1 \le i \le N_x, 1 \le j \le N_y} |a_{ij}|, \tag{10}$$

respectively. The discretizations (5) and (6) are conservative, that is,

$$\sum_{i=1}^{N_x}\sum_{j=1}^{N_y}\phi_{ij}^{n+1}=\sum_{i=1}^{N_x}\sum_{j=1}^{N_y}\phi_{ij}^{n}. \tag{11}$$

To show this conservation property, we take the summation of Equation (5)

$$\begin{aligned}\sum_{i=1}^{N_x}\sum_{j=1}^{N_y}\frac{\phi_{ij}^{n+1}-\phi_{ij}^{n}}{\Delta t} &= \sum_{i=1}^{N_x}\sum_{j=1}^{N_y}\Delta_h\mu_{ij}^{n+1} \\ &= \sum_{j=1}^{N_y}\left(\frac{\mu_{N_x+1,j}^{n+1}-\mu_{N_xj}^{n+1}}{h^2}-\frac{\mu_{1j}^{n+1}-\mu_{0j}^{n+1}}{h^2}\right) \\ &\quad+\sum_{i=1}^{N_x}\left(\frac{\mu_{i,N_y+1}^{n+1}-\mu_{iN_y}^{n+1}}{h^2}-\frac{\mu_{i1}^{n+1}-\mu_{i0}^{n+1}}{h^2}\right)=0.\end{aligned} \tag{12}$$

Here, we used the homogenous Neumann boundary conditions (7) and (8). Therefore, Equation (11) holds. We define the discrete energy functional as

$$\begin{aligned}\mathcal{E}^h(\phi^n) &= h^2\sum_{i=1}^{N_x}\sum_{j=1}^{N_y}F(\phi_{ij}^n) \\ &\quad+\frac{\epsilon^2}{2}\sum_{j=1}^{N_y}\left(\frac{(\phi_{1j}^n-\phi_{0j}^n)^2}{2}+\sum_{i=1}^{N_x-1}(\phi_{i+1,j}^n-\phi_{ij}^n)^2+\frac{(\phi_{N_x+1,j}^n-\phi_{N_xj}^n)^2}{2}\right) \\ &\quad+\frac{\epsilon^2}{2}\sum_{i=1}^{N_x}\left(\frac{(\phi_{i1}^n-\phi_{i0}^n)^2}{2}+\sum_{j=1}^{N_y-1}(\phi_{i,j+1}^n-\phi_{ij}^n)^2+\frac{(\phi_{i,N_y+1}^n-\phi_{iN_y}^n)^2}{2}\right) \\ &= h^2\sum_{i=1}^{N_x}\sum_{j=1}^{N_y}F(\phi_{ij}^n)+\frac{\epsilon^2}{2}\sum_{j=1}^{N_y}\sum_{i=1}^{N_x-1}(\phi_{i+1,j}^n-\phi_{ij}^n)^2+\frac{\epsilon^2}{2}\sum_{i=1}^{N_x}\sum_{j=1}^{N_y-1}(\phi_{i,j+1}^n-\phi_{ij}^n)^2,\end{aligned} \tag{13}$$

where we used the homogenous Neumann boundary conditions (7) and (8). We also define the discrete total mass as

$$\mathcal{M}^h(\phi^n)=\sum_{i=1}^{N_x}\sum_{j=1}^{N_y}\phi_{ij}^n h^2. \tag{14}$$

Then, the unconditionally gradient stable scheme satisfies the reduction in the discrete total energy [24]:

$$\mathcal{E}^h(\phi^{n+1})\le\mathcal{E}^h(\phi^n), \tag{15}$$

which implies the pointwise boundedness of the numerical solution:

$$\|\phi^n\|_\infty\le\sqrt{1+2\sqrt{\mathcal{E}^h(\boldsymbol{\phi}^0)/h^2}}\quad\text{for all } n. \tag{16}$$

The proof of Equation (16) can be found in Reference [25]. We provide the proof herein for the sake of completeness. We show that a constant $K$ exists for all $n$ values that satisfy the following inequality:

$$\|\phi^n\|_\infty\le K. \tag{17}$$

Let us assume that there is an integer $n_K$ that is dependent on $K$ such that $\|\phi^{n_K}\|_\infty > K$ for any $K$. Then, $\phi_{ij}^{n_K}$ exists such that $|\phi_{ij}^{n_K}| > K$. Let $K$ be the largest solution of $\mathcal{E}^h(\phi^0) = h^2F(K)$, that is, $K = \sqrt{1+2\sqrt{\mathcal{E}^h(\phi^0)/h^2}}$. We then have

$$\mathcal{E}^h(\phi^0) = h^2F(K) < h^2F(|\phi_{ij}^{n_K}|) \le \mathcal{E}^h(\phi^{n_K}) \le \mathcal{E}^h(\phi^0), \tag{18}$$

where we utilize the fact that the total energy is decreasing and $F(\phi)$ is a strictly increasing function on $(K,\infty)$. Equation (18) leads to a contradiction. Therefore, Equation (17) should be satisfied.

### 2.2. Multigrid V-Cycle Algorithm

We use the nonlinear full approximation storage (FAS) multigrid method to solve the nonlinear discrete systems (5) and (6). For simplicity, we define the discrete domains, $\Omega_2$, $\Omega_1$, and $\Omega_0$, which represent a hierarchy of meshes ($\Omega_2$, $\Omega_1$, and $\Omega_0$) created by successively coarsening the original mesh $\Omega_2$, as shown in Figure 2.

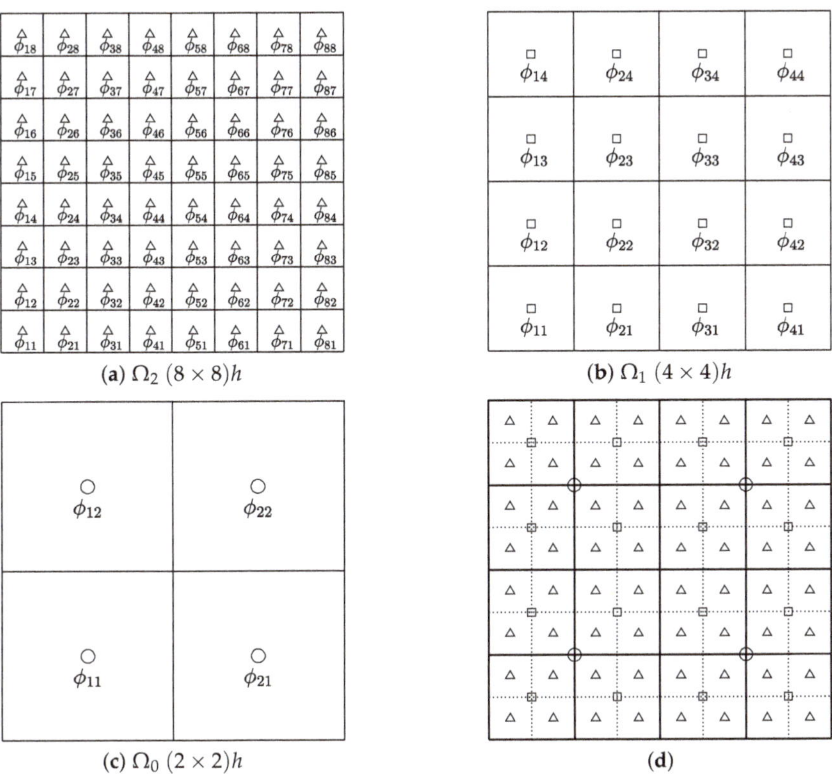

**Figure 2.** (**a–c**) represent a sequence of coarse grids starting with $h = L/N_x$. (**d**) depicts a composition of grids, $\Omega_2$, $\Omega_1$ and $\Omega_0$.

We summarize here the nonlinear multigrid method for solving the discrete CH system as follows: First, let us rewrite Equations (5) and (6) as

$$NSO(\phi^{n+1}, \mu^{n+1}) = (\xi^n, \psi^n),$$

where the linear operator $NSO$ is defined as

$$NSO(\phi^{n+1}, \mu^{n+1}) = \left(\phi^{n+1}/\Delta t - \Delta_h \mu^{n+1},\ \mu^{n+1} - (\phi^{n+1})^3 + \epsilon^2 \Delta_h \phi^{n+1}\right),$$

and the source term is denoted by

$$(\xi^n, \psi^n) = (\phi^n/\Delta t,\ -\phi^n). \tag{19}$$

Next, we describe the multigrid method, which includes the pre-smoothing, coarse grid correction and post-smoothing steps. We denote a mesh grid $\Omega_k$ as the discrete domain for each multigrid level $k$. Note that a mesh grid $\Omega_k$ contains $2^k \times 2^k$ grid points. Let $k_{min}$ be the coarsest multigrid level. We now introduce the SMOOTH and V-cycle functions. Given the number $\nu_1$ of pre-smoothing and $\nu_2$ of post-smoothing relaxation sweeps, the V-cycle is used as an iteration step in the multigrid method.

*FAS multigrid cycle*

Now, we define the FAScycle:

$$\{\phi_k^{m+1}, \mu_k^{m+1}\} = FAScycle(k, \phi_k^m, \mu_k^m, NSO_k, \xi_k^n, \psi_k^n, \beta).$$

In other words, $\{\phi_k^m, \mu_k^m\}$ and $\{\phi_k^{m+1}, \mu_k^{m+1}\}$ are the approximations of $\phi^{n+1}(x_i, y_j)$ and $\mu^{n+1}(x_i, y_j)$ before and after an FAScycle, respectively. Here, $\phi_k^0 = \phi_k^n$ and $\mu_k^0 = \mu_k^n$.

*(1) Pre-smoothing*

$$\{\bar{\phi}_k^m,\ \bar{\mu}_k^m\} = SMOOTH^{\nu_1}(\phi_k^m, \mu_k^m, NSO_k, \xi_k^n, \psi_k^n),$$

which represents $\nu_1$ smoothing steps with the initial approximations $\phi_k^m, \mu_k^m$, source terms $\xi_k^n, \psi_k^n$ and the $SMOOTH$ relaxation operator to obtain the approximations $\bar{\phi}_k^m,\ \bar{\mu}_k^m$. One $SMOOTH$ relaxation operator step consists of solving the systems (22) and (23), given as follows by $2 \times 2$ matrix inversion for each $i$ and $j$. Here, we derive the smoothing operator in two dimensions. Rewriting Equation (5), we obtain:

$$\frac{\phi_{ij}^{n+1}}{\Delta t} + \frac{4\mu_{ij}^{n+1}}{h^2} = \xi_{ij}^n + \frac{\mu_{i+1,j}^{n+1} + \mu_{i-1,j}^{n+1} + \mu_{i,j+1}^{n+1} + \mu_{i,j-1}^{n+1}}{h^2}. \tag{20}$$

Because $(\phi_{ij}^{n+1})^3$ is nonlinear with respect to $\phi_{ij}^{n+1}$, we linearize $(\phi_{ij}^{n+1})^3$ at $\phi_{ij}^m$, that is,

$$(\phi_{ij}^{n+1})^3 \approx (\phi_{ij}^m)^3 + 3(\phi_{ij}^m)^2(\phi_{ij}^{n+1} - \phi_{ij}^m).$$

After substituting of this into (6), we obtain

$$-\left(\frac{4\epsilon^2}{h^2} + 3(\phi_{ij}^m)^2\right)\phi_{ij}^{n+1} + \mu_{ij}^{n+1} = \psi_{ij}^n - 2(\phi_{ij}^m)^3 - \frac{\epsilon^2}{h^2}(\phi_{i+1,j}^{n+1} + \phi_{i-1,j}^{n+1} + \phi_{i,j+1}^{n+1} + \phi_{i,j-1}^{n+1}). \tag{21}$$

Next, we replace $\phi_{\alpha,\beta}^{n+1}$ and $\mu_{\alpha,\beta}^{n+1}$ in Equations (20) and (21) with $\bar{\phi}_{\alpha,\beta}^m$ and $\bar{\mu}_{\alpha,\beta}^m$ for $\alpha \le i$ and $\beta \le j$, otherwise with $\phi_{\alpha,\beta}^m$ and $\mu_{\alpha,\beta}^m$, that is,

$$\frac{\bar{\phi}_{ij}^m}{\Delta t} + \frac{4\bar{\mu}_{ij}^m}{h^2} = \xi_{ij}^n + \frac{\mu_{i+1,j}^m + \bar{\mu}_{i-1,j}^m + \mu_{i,j+1}^m + \bar{\mu}_{i,j-1}^m}{h^2}, \tag{22}$$

$$-\left(\frac{4\epsilon^2}{h^2} + 3(\phi_{ij}^m)^2\right)\bar{\phi}_{ij}^m + \bar{\mu}_{ij}^m = \psi_{ij}^n - 2(\phi_{ij}^m)^3 - \frac{\epsilon^2}{h^2}(\phi_{i+1,j}^m + \bar{\phi}_{i-1,j}^m + \phi_{i,j+1}^m + \bar{\phi}_{i,j-1}^m). \tag{23}$$

*(2) Compute the defect*

$$(\bar{d_1}^m{}_k, \bar{d_2}^m{}_k) = (\xi_k^n, \psi_k^n) - NSO_k(\bar{\phi}_k^m, \bar{\mu}_k^m).$$

*(3) Restrict the defect and* $\{\bar{\phi}_k^m,\ \bar{\mu}_k^m\}$

$$(\bar{d}_{1k-1}^{\ m}, \bar{d}_{2k-1}^{\ m}) = I_k^{k-1}(\bar{d}_{1k}^{\ m}, \bar{d}_{2k}^{\ m}).$$

The restriction operator $I_k^{k-1}$ maps $k$-level functions to $(k-1)$-level functions.

$$\begin{aligned} d_{k-1}(x_i, y_j) = I_k^{k-1} d_k(x_i, y_j) \ = \ & \frac{1}{4}[d_k(x_{i-\frac{1}{2}}, y_{j-\frac{1}{2}}) + d_k(x_{i-\frac{1}{2}}, y_{j+\frac{1}{2}}) \\ & + d_k(x_{i+\frac{1}{2}}, y_{j-\frac{1}{2}}) + d_k(x_{i+\frac{1}{2}}, y_{j+\frac{1}{2}})]. \end{aligned}$$

*(4) Compute the right-hand side*

$$(\xi_{k-1}^n, \psi_{k-1}^n) = (\bar{d_1}^m{}_{k-1}, \bar{d_2}^m{}_{k-1}) + NSO_{k-1}(\bar{\phi}_{k-1}^m, \bar{\mu}_{k-1}^m).$$

*(5) Compute an approximate solution* $\{\hat{\phi}_{k-1}^m,\ \hat{\mu}_{k-1}^m\}$ *of the coarse grid equation on* $\Omega_{k-1}$, *that is,*

$$NSO_{k-1}(\phi_{k-1}^m, \mu_{k-1}^m) = (\xi_{k-1}^n, \psi_{k-1}^n). \tag{24}$$

If $k = 1$, we explicitly invert the $2 \times 2$ matrix to obtain the solution. If $k > 1$, we solve Equation (24) by performing a FAS $k$-grid cycle using $\{\bar{\phi}_{k-1}^m,\ \bar{\mu}_{k-1}^m\}$ as the initial approximation:

$$\{\hat{\phi}_{k-1}^m, \hat{\mu}_{k-1}^m\} = \text{FAScycle}(k-1, \bar{\phi}_{k-1}^m, \bar{\mu}_{k-1}^m, NSO_{k-1}, \xi_{k-1}^n, \psi_{k-1}^n, \beta).$$

*(6) Compute the coarse grid correction (CGC):*

$$\hat{v}_{1k-1}^m = \hat{\phi}_{k-1}^m - \bar{\phi}_{k-1}^m, \quad \hat{v}_{2k-1}^m = \hat{\mu}_{k-1}^m - \bar{\mu}_{k-1}^m.$$

*(7) Interpolate the correction:*

$$\hat{v}_{1k}^m = I_{k-1}^k \hat{v}_{1k-1}^m, \quad \hat{v}_{2k}^m = I_{k-1}^k \hat{v}_{2k-1}^m.$$

Here, the coarse values are simply transferred to the four nearby fine grid points, that is, $v_k(x_i, y_j) = I_{k-1}^k v_{k-1}(x_i, y_j) = v_{k-1}(x_{i+\frac{1}{2}}, y_{j+\frac{1}{2}})$ for $i$ and $j$ odd-numbered integers.

*(8) Compute the corrected approximation on* $\Omega_k$

$$\phi_k^{m,\text{ after CGC}} = \bar{\phi}_k^m + \hat{v}_{1k}^{\ m}, \quad \mu_k^{m,\text{ after CGC}} = \bar{\mu}_k^m + \hat{v}_{2k}^{\ m}.$$

*(9) Post-smoothing*

$$\{\phi_k^{m+1}, \mu_k^m\} = SMOOTH^{\nu_2}(\phi_k^{m,\text{ after CGC}}, \mu_k^{m,\text{ after CGC}}, NSO_k, \xi_k^n, \psi_k^n).$$

This completes the description of the nonlinear FAScycle. One FAScycle step stops if the consequent error $\|\phi^{n+1,m+1} - \phi^{n+1,m}\|_\infty$ is less than a given tolerance *tol*. The two-grid V-cycle is illustrated in Figure 3.

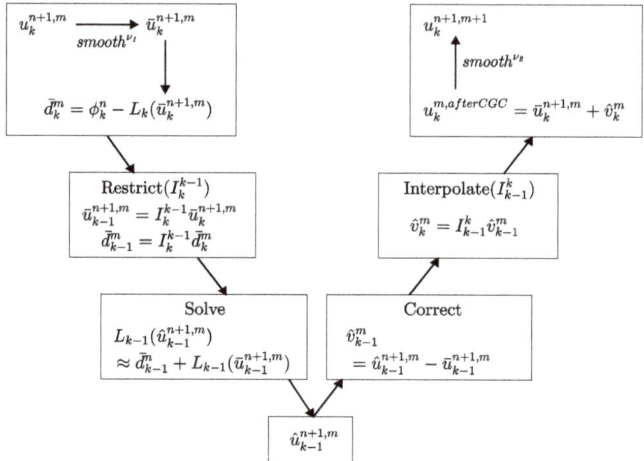


**Figure 3.** Multigrid two-grid V-cycle method.

Further Numerical Schemes for the CH Equation

Previous studies have described the numerical solution of the CH equation with a variable mobility [19], the adaptive mesh refinement technique [26,27], the Neumann boundary condition in complex domains [20], the Dirichlet boundary conditions in complex domains [28], contact angle boundary [29], parallel multigrid method [30] and fourth-order compact scheme [31].

## 3. Numerical Experiments

In numerical experiments, we consider an equilibrium solution $\phi(x,\infty) = \tanh(x/\sqrt{2}\epsilon)$ for the CH equation on the one-dimensional infinite domain $\Omega = (-\infty,\infty)$. In other words, $\phi(x,\infty)$ satisfies $\mu(\phi(x,\infty)) = F'(\phi(x,\infty)) - \epsilon^2\phi_{xx}(x,\infty) = 0$ and is an equilibrium solution. Then, across the interfacial regions, $\phi$ varies from $-0.9$ to $0.9$ over a distance of approximately $\xi = 2\sqrt{2}\epsilon\tanh^{-1}(0.9)$ (see Figure 4). Therefore, if we want this value to be approximately $mh$, the $\epsilon$ value can be taken as $\epsilon = \epsilon_m = mh/[2\sqrt{2}\tanh^{-1}(0.9)]$ [32].

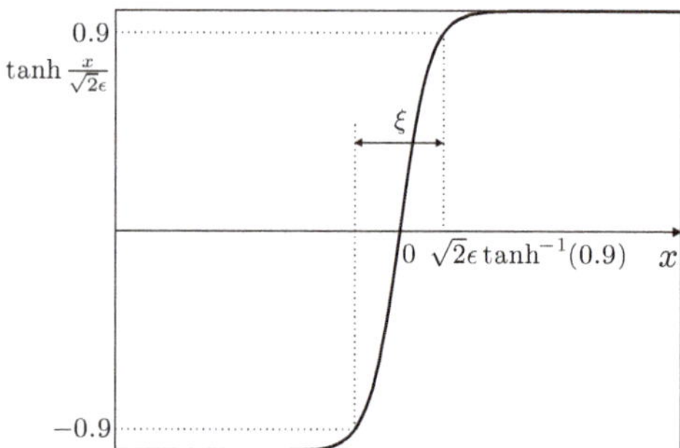


**Figure 4.** Concentration field varying from $-0.9$ to $0.9$ over a distance of approximately $\xi = 2\sqrt{2}\epsilon\tanh^{-1}(0.9)$.

All computational simulations described in this section are performed on an Intel Core i5-6400 CPU @ 2.70 GHz with 4 GB of RAM.

### 3.1. Phase Separation

For the first numerical test, we consider spinodal decomposition in binary alloys. This decomposition is a process by which a mixture of binary materials separates into distinct regions with different material concentrations [2]. Figure 5a–c show snapshots of the phase-field $\phi$ at $t = 100\Delta t$, $200\Delta t$ and $1000\Delta t$, respectively. The initial condition is $\phi(x, y, 0) = 0.1(1 - 2\text{rand}(x, y))$ on $\Omega = (0, 1) \times (0, 1)$, where $\text{rand}(x, y)$ is a random value between 0 and 1. The parameters $\epsilon = \epsilon_4$, $h = 1/64$, $\Delta t = 0.1h^2$ and a tolerance of $tol = 1.0 \times 10^{-10}$ are used.

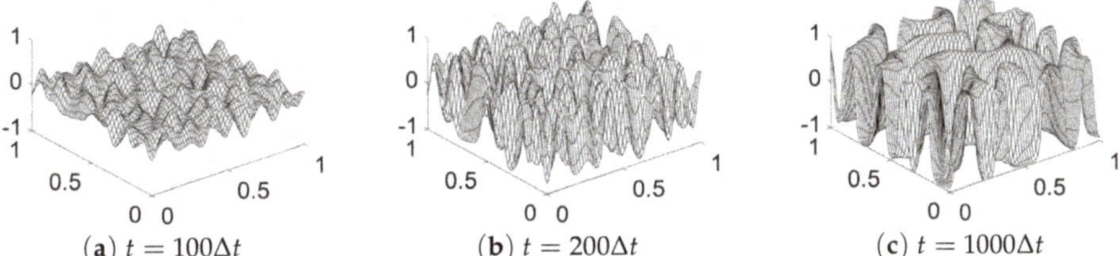


**Figure 5.** Snapshots of the phase-field $\phi$ at **(a)** $t = 100\Delta t$, **(b)** $t = 200\Delta t$ and **(c)** $t = 1000\Delta t$. Here, $\epsilon = \epsilon_4$, $h = 1/64$ and $\Delta t = 0.1h^2$ are used.

### 3.2. Non-Increase in Discrete Energy and Mass Conservation

Figure 6 shows the time evolution of the normalized discrete total energy $\mathcal{E}^h(\phi^n)/\mathcal{E}^h(\phi^0)$ (solid line) and the average mass $\mathcal{M}^h(\phi^n)/(h^2 N_x N_y)$ (diamond) of the numerical solutions with the initial state (25) on $\Omega = (0, 1) \times (0, 1)$.

$$\phi(x, y, 0) = 0.1(1 - 2\text{rand}(x, y)), \tag{25}$$

where $\text{rand}(x, y)$ is a random value between 0 and 1.

We use the simulation parameters, $\epsilon = \epsilon_4$, $h = 1/64$, $\Delta t = 0.1h^2$ and $tol = 1.0 \times 10^{-10}$. The energy is non-increasing and the average concentration is conserved. These numerical results agree well with the total energy dissipation property (3) and the conservation property (4). The inscribed small figures are the concentration fields at the indicated times.

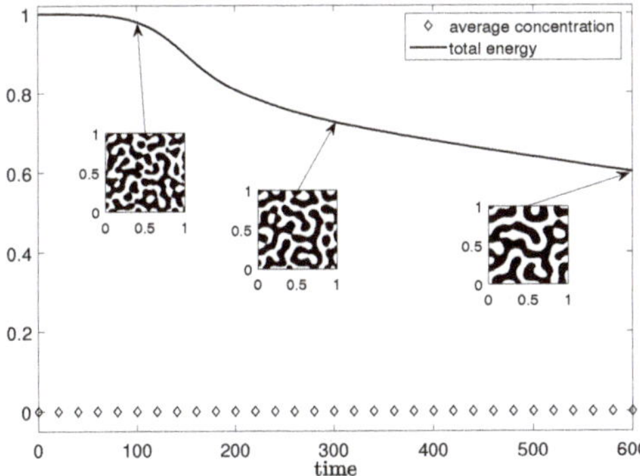


**Figure 6.** Normalized discrete total energy $\mathcal{E}^h(\phi^n)/\mathcal{E}^h(\phi^0)$ (solid line) and average concentration $\mathcal{M}^h(\phi^n)/(h^2 N_x N_y)$ (diamond line) of the numerical solutions with the initial state (25).

### 3.3. Convergence Test

We consider the convergence of the Frobenius norm with a scaling of the residual error with respect to the grid size. The initial condition on the domain $\Omega = (0,1) \times (0,1)$ is given as

$$\phi(x,y,0) = 0.1\cos(\pi x)\cos(\pi y). \tag{26}$$

We fix $\epsilon = 0.06$, $\Delta t = 0.01$ and $tol = 1.0 \times 10^{-15}$. Here, we use the $V(2,2)$ scheme with a Gauss–Seidel relaxation, where $(2,2)$ indicates 2 pre- and 2 post-correction relaxation sweeps. We define the residual after $m$ $V$-cycles as

$$r_{ij}^{m} = \Delta_h \mu_{ij}^{n+1,m} - \frac{\phi_{ij}^{n+1,m} - \phi_{ij}^{n}}{\Delta t}. \tag{27}$$

Table 1 shows the residual norm $\|r^m\|_F$ after each $V$-cycle. Because no closed-form analytical solution exists for this problem, we define the Frobenius norm with a scaling of the residual error $\|r^m\|_F = \|\Delta_h \mu^{n+1,m} - (\phi^{n+1,m} - \phi^n)/\Delta t\|_F$. The grid sizes are set as $32 \times 32$, $64 \times 64$ and $128 \times 128$. The error norms and ratios of residual between successive V-cycle are shown in Table 1. As we have expected, the residual error decreases successively along with the V-cycle. The sharp increase in the residual norm ratio during the last few cycles reflects the fact that the numerical approximation is already accurate to near machine precision.

**Table 1.** Error and convergence results for various grid spaces.

| Mesh Size | $32 \times 32$ | | $64 \times 64$ | | $128 \times 128$ | |
|---|---|---|---|---|---|---|
| **V-Cycle** | $\|r\|_F$ | **Ratio** | $\|r\|_F$ | **Ratio** | $\|r\|_F$ | **Ratio** |
| 1 | $5.28 \times 10^{-2}$ | | $6.83 \times 10^{-2}$ | | $9.10 \times 10^{-2}$ | |
| 2 | $2.41 \times 10^{-3}$ | 0.05 | $3.42 \times 10^{-3}$ | 0.05 | $4.30 \times 10^{-3}$ | 0.05 |
| 3 | $1.15 \times 10^{-4}$ | 0.05 | $1.56 \times 10^{-4}$ | 0.05 | $2.09 \times 10^{-4}$ | 0.05 |
| 4 | $6.54 \times 10^{-6}$ | 0.06 | $7.25 \times 10^{-6}$ | 0.05 | $8.69 \times 10^{-6}$ | 0.04 |
| 5 | $4.23 \times 10^{-7}$ | 0.06 | $4.23 \times 10^{-7}$ | 0.06 | $4.66 \times 10^{-7}$ | 0.05 |
| 6 | $2.80 \times 10^{-8}$ | 0.07 | $2.65 \times 10^{-8}$ | 0.06 | $2.90 \times 10^{-8}$ | 0.06 |
| 7 | $1.83 \times 10^{-9}$ | 0.07 | $1.58 \times 10^{-9}$ | 0.06 | $1.63 \times 10^{-9}$ | 0.06 |
| 8 | $1.23 \times 10^{-10}$ | 0.07 | $1.01 \times 10^{-10}$ | 0.06 | $1.01 \times 10^{-10}$ | 0.06 |
| 9 | $8.33 \times 10^{-12}$ | 0.07 | $6.75 \times 10^{-12}$ | 0.07 | $6.83 \times 10^{-12}$ | 0.07 |
| 10 | $5.46 \times 10^{-13}$ | 0.07 | $4.24 \times 10^{-13}$ | 0.06 | $4.42 \times 10^{-13}$ | 0.06 |
| 11 | $3.70 \times 10^{-14}$ | 0.07 | $4.90 \times 10^{-14}$ | 0.12 | $1.69 \times 10^{-13}$ | 0.38 |
| 12 | $1.10 \times 10^{-14}$ | 0.30 | $4.10 \times 10^{-14}$ | 0.84 | $1.66 \times 10^{-13}$ | 0.98 |

### 3.4. Effect of Tolerance

The effect of multigrid tolerance is related to the average mass convergence. We set the initial condition $\phi(x,y,0) = 0.1\cos(\pi x)\cos(\pi y)$ on $\Omega = (0,1) \times (0,1)$ with tolerance, $tol = 1.0 \times 10^{-1}$, $1.0 \times 10^{-2}$, $1.0 \times 10^{-3}$, and $1.0 \times 10^{-10}$ to investigate the relationship between the mass convergence and $tol$. We use the simulation parameters $\epsilon = \epsilon_4$, SMOOTH relaxation = 2, $h = 1/32$, $\Delta t = 1/32$ and mesh size $32 \times 32$. To compare the theoretical value (solid line) with the computational value $tol = 1.0 \times 10^{-1}$ (dotted line), $tol = 1.0 \times 10^{-2}$ (dash-dot line), $tol = 1.0 \times 10^{-3}$ (dashed line) and $tol = 1.0 \times 10^{-10}$ (square), we set the interval of average mass from $-0.0019$ to $0.0023$. In Figure 7, the average mass $\mathcal{M}^h(\phi^n)/(h^2 N_x N_y)$ gradually converges to a theoretical value with the decrease in tolerance. In addition, comparing the results of $tol = 1.0 \times 10^{-1}$, $1.0 \times 10^{-2}$, $1.0 \times 10^{-3}$ and $1.0 \times 10^{-10}$, we observe that the average mass become nearly convergent for $tol = 1.0 \times 10^{-10}$.

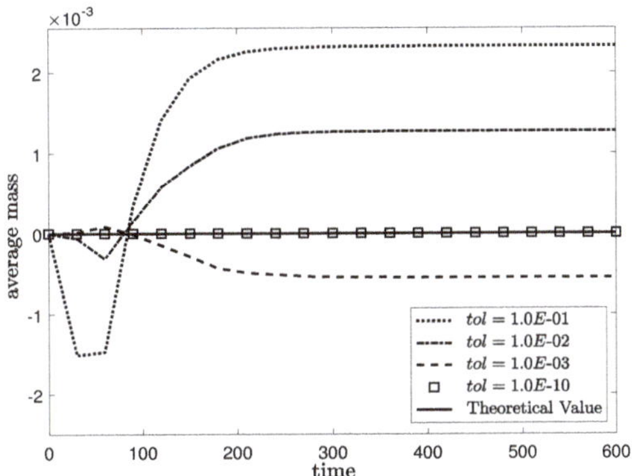


**Figure 7.** Average mass $\mathcal{M}^h(\phi^n)/(h^2N_xN_y)$ of the numerical solutions in various values of tolerance. Here, the theoretical value (solid line), $tol = 1.0 \times 10^{-1}$ (dotted line), $tol = 1.0 \times 10^{-2}$ (dash-dot line), $tol = 1.0 \times 10^{-3}$ (dashed line) and $tol = 1.0 \times 10^{-10}$ (square).

### 3.5. Effects of the Smooth Relaxation Numbers $\nu_1$ and $\nu_2$

We investigate the effects of the SMOOTH relaxation numbers $\nu_1$ (pre-relaxation) and $\nu_2$ (post-relaxation) on the CPU time. In this test, we perform a numerical simulation with the initial condition $\phi(x,y,0) = 0.1\cos(\pi x)\cos(\pi y)$ on $\Omega = (0,1)\times(0,1)$, $h = 1/128$, $\epsilon_4$, $\Delta t = 0.1h^2$ and $tol = 1.0 \times 10^{-10}$. Table 2 lists the average CPU times and average numbers of V-cycles for various pre- and post-relaxation numbers after 100 time steps. The relaxation numbers are rounded off to the nearest integer. Figure 8 shows the average CPU times for different pre- and post-relaxation numbers. We observe that the average CPU time is the lowest when the numbers of pre- and post-relaxation iterations are $\nu_1 = 2$ and $\nu_2 = 4$, respectively.

**Table 2.** Average CPU times and average numbers of V-cycles (given in parentheses) for various pre- and post-relaxation numbers after 100 time steps.

| $\nu_2$ \ $\nu_1$ | 1 | 2 | 3 | 4 | 5 |
|---|---|---|---|---|---|
| 1 | 0.075(9) | 0.075(7) | 0.081(6) | 0.077(5) | 0.089(5) |
| 2 | 0.075(7) | 0.079(6) | 0.090(5) | 0.089(5) | 0.081(4) |
| 3 | 0.079(6) | 0.082(5) | 0.089(5) | 0.081(4) | 0.090(4) |
| 4 | 0.078(5) | 0.075(4) | 0.081(4) | 0.090(4) | 0.075(3) |
| 5 | 0.072(4) | 0.081(4) | 0.090(4) | 0.075(3) | 0.082(3) |

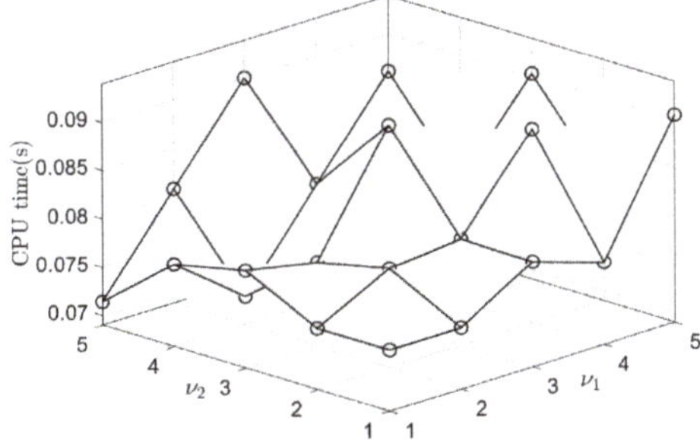


**Figure 8.** Average CPU time for different pre- and post-relaxation numbers after 100 time steps.

Next, we investigate the effect of SMOOTH relaxation numbers on the finest multigrid level. We perform a numerical simulation with $\epsilon = 0.0038$ and $\Delta t = 1.0 \times 10^{-7}$. The SMOOTH relaxation numbers $\nu_0^1$ and $\nu_0^2$ (on the finest multigrid level) are taken to be 1, 2, 3, 4, 5, 6, 7, 8, 9 and 10. In addition, other multigrid levels $\nu_k^1 = \nu_k^2$ are 1, 2, and 3 in the 128 × 128 and 512 × 512 mesh sizes. The other parameters are the same as those used previously. Table 3 shows the variations in average CPU time for different relaxation numbers with a 128 × 128 mesh size. Figure 9 illustrates the results in Table 3.

**Table 3.** Average CPU times for various relaxation numbers on finest multigrid level $(\nu_0^1, \nu_0^2)$ after 10 time steps. In other grids, $\nu_k^1 = \nu_k^2, (1 \leq k)$ relaxation number is fixed at 1,2 and 3.

| **128 × 128** | | $\nu_k^1 = \nu_k^2$ | |
|---|---|---|---|
| $\nu_0^1 = \nu_0^2$ | **1** | **2** | **3** |
| 1 | 0.043 | 0.049 | 0.054 |
| 2 | 0.040 | 0.045 | 0.048 |
| 3 | 0.027 | 0.029 | 0.031 |
| 4 | 0.032 | 0.034 | 0.037 |
| 5 | 0.038 | 0.041 | 0.042 |
| 6 | 0.044 | 0.047 | 0.048 |
| 7 | 0.050 | 0.053 | 0.054 |
| 8 | 0.055 | 0.058 | 0.060 |
| 9 | 0.062 | 0.065 | 0.065 |
| 10 | 0.068 | 0.070 | 0.072 |

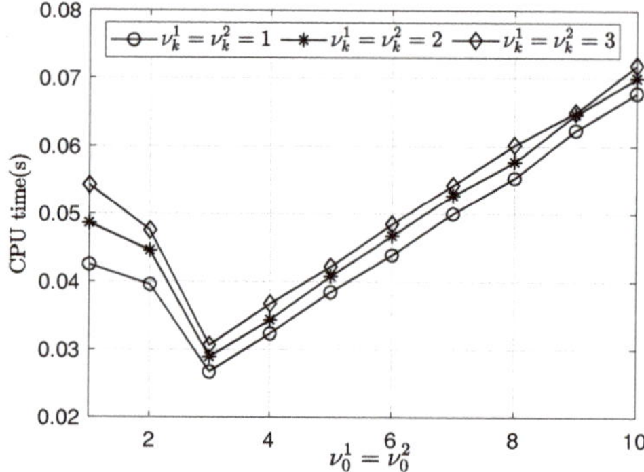


**Figure 9.** Average CPU times for various relaxation numbers on finest multigrid level $(\nu_0^1, \nu_0^2)$ and fixed relaxation number 1,2 and 3 on other grids with a 128 × 128 mesh size.

Table 4 lists the variations in average CPU time for different relaxation numbers with a 512 × 512 mesh size. Figure 10 illustrates the results in Table 4.

**Table 4.** Average CPU times for various relaxation numbers on finest multigrid level $(\nu_0^1, \nu_0^2)$ after 10 time steps. In other grids, the $\nu_k^1 = \nu_k^2, (1 \leq k)$ relaxation number is fixed at 1,2 and 3.

| **512 × 512** | | $\nu_k^1 = \nu_k^2$ | |
|---|---|---|---|
| $\nu_0^1 = \nu_0^2$ | **1** | **2** | **3** |
| 1 | 2.182 | 2.506 | 2.750 |
| 2 | 2.030 | 2.208 | 2.397 |
| 3 | 1.726 | 1.850 | 1.982 |
| 4 | 2.096 | 2.227 | 2.351 |
| 5 | 1.872 | 1.943 | 2.047 |
| 6 | 2.151 | 2.236 | 2.330 |
| 7 | 2.426 | 2.514 | 2.617 |
| 8 | 1.817 | 1.877 | 1.944 |
| 9 | 2.002 | 2.066 | 2.118 |
| 10 | 2.197 | 2.252 | 2.295 |

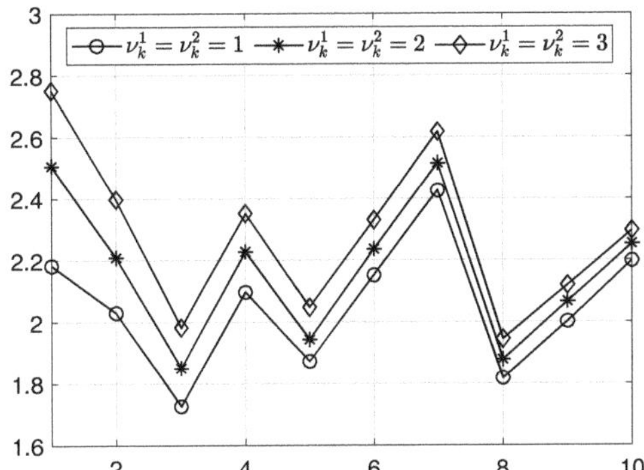


**Figure 10.** Average CPU times for various relaxation numbers on finest multigrid level $(\nu_0^1, \nu_0^2)$ and fixed relaxation numbers 1, 2 and 3 on other grids with a $512 \times 512$ mesh size.

## 3.6. Effect Of V-Cycle

Next, we investigate the effect of V-cycle by changing the multigrid levels. In this test, we use the parameters $\Delta t = 0.01$, $\epsilon = 0.06$, SMOOTH relaxation $= 2$, $tol = 1.0 \times 10^{-10}$ and $h = 1/128$ on $\Omega = (0,1) \times (0,1)$ with a initial condition $\phi(x,y,0) = 0.1\cos(\pi x)\cos(\pi y)$. The highest number of V-cycles is taken to be 10000. We use $level_2$, $level_3$, $level_4$, $level_5$, $level_6$ and $level_7$ in a single time step as examples to illustrate the effect of the V-cycle. We calculate the CPU time for each level after $100\Delta t$, as listed in Table 5.

**Table 5.** Numbers of multigrid levels and CPU times required until tolerance $\leq 1.0 \times 10^{-10}$. Here, different levels are used.

| **Level** | **2** | **3** | **4** | **5** | **6** | **7** |
|---|---|---|---|---|---|---|
| CPU time(s) | 1271.437 | 342.906 | 103.032 | 66.422 | 66.172 | 65.218 |

The number of the multigrid level and CPU time shown in Figure 11 indicate that a greater number of the multigrid level leads to a obvious decrease in CPU time. It is important to select an appropriate multigrid level for a specific mesh size.

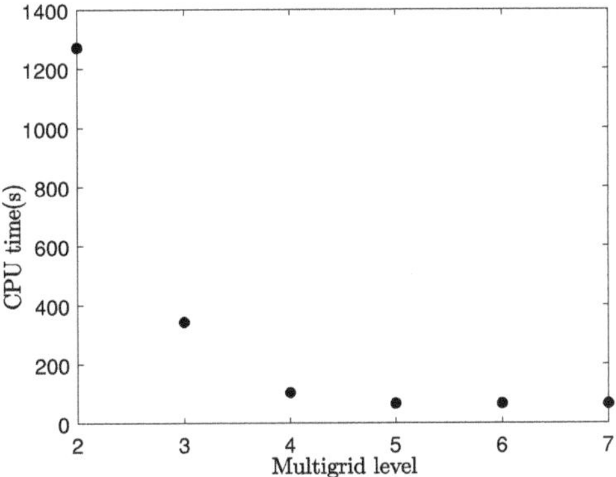


**Figure 11.** Required CPU times in various multigrid levels after 100 time steps.

## 3.7. Comparison between Gauss–Seidel and Multigrid Algorithms

We compare the average CPU times required to perform the Gauss–Seidel algorithm and multigrid algorithm. In this test, the initial condition is taken to be $\phi(x,y,0) = \cos(\pi x)\cos(\pi y)$. The following parameters are used—$\Delta t = 0.01$, $T = 10\Delta t$, the SMOOTH relaxation $= 2$ and $\epsilon = 0.06$. The highest number of V-cycle is taken to be 10000. The mesh sizes are $32 \times 32$, $64 \times 64$ and $128 \times 128$. The tolerances are $1.0 \times 10^{-3}$, $1.0 \times 10^{-4}$ and $1.0 \times 10^{-5}$. Table 6 shows the average CPU times for these two methods after 10 time steps. We observe that the multigrid method require less CPU time than the Gauss–Seidel method does.

**Table 6.** Average CPU times for Gauss–Seidel and multigrid algorithms with different tolerances after 10 time steps.

| Mesh Size | *tol* | Gauss–Seidel | Multigrid |
|---|---|---|---|
| $32 \times 32$ | $1.0 \times 10^{-3}$ | 0.468 | 0.046 |
| | $1.0 \times 10^{-4}$ | 0.610 | 0.063 |
| | $1.0 \times 10^{-5}$ | 0.735 | 0.062 |
| $64 \times 64$ | $1.0 \times 10^{-3}$ | 7.046 | 0.203 |
| | $1.0 \times 10^{-4}$ | 8.984 | 0.234 |
| | $1.0 \times 10^{-5}$ | 11.360 | 0.266 |
| $128 \times 128$ | $1.0 \times 10^{-3}$ | 109.093 | 0.844 |
| | $1.0 \times 10^{-4}$ | 140.219 | 0.906 |
| | $1.0 \times 10^{-5}$ | 174.922 | 1.078 |

### 3.8. Effects of *tol* and $\Delta t$ on the V-Cycle

In this test, we study the effects of *tol* and $\Delta t$ on the V-cycle with the initial condition being $\phi(x, y, 0) = 0.1\cos(\pi x)\cos(\pi y)$, $h = 1/128$, $\epsilon = 0.06$. The highest number of the V-cycle is taken to be 10,000. Table 7 shows the number of V-cycle for various *tol* and $\Delta t$ after a single time step. We can find that lower values of *tol* lead to an increase in the V-cycle. For different values of *tol*, it is essential to choose an appropriate $\Delta t$ to reduce the number of V-cycle.

**Table 7.** Numbers of V-cycles for various *tol* and $\Delta t$ after a single time step.

| *tol* \ $\Delta t$ | $10^{-7}h^2$ | $10^{-6}h^2$ | $10^{-5}h^2$ | $10^{-4}h^2$ | $10^{-3}h^2$ | $10^{-2}h^2$ | $10^{-1}h^2$ |
|---|---|---|---|---|---|---|---|
| $1.0 \times 10^{-5}$ | 2 | 2 | 3 | 4 | 6 | 8 | 8 |
| $1.0 \times 10^{-6}$ | 10,000 | 2 | 3 | 4 | 7 | 9 | 9 |
| $1.0 \times 10^{-7}$ | 10,000 | 10,000 | 3 | 5 | 8 | 10 | 10 |
| $1.0 \times 10^{-8}$ | 10,000 | 10,000 | 10,000 | 5 | 9 | 11 | 12 |
| $1.0 \times 10^{-9}$ | 10,000 | 10,000 | 10,000 | 10,000 | 10 | 13 | 13 |
| $1.0 \times 10^{-10}$ | 10,000 | 10,000 | 10,000 | 10,000 | 10,000 | 14 | 15 |
| *tol* \ $\Delta t$ | $h^2$ | $10h^2$ | $10^2h^2$ | $10^3h^2$ | $10^4h^2$ | $10^5h^2$ | $10^6h^2$ |
| $1.0 \times 10^{-5}$ | 8 | 7 | 8 | 8 | 9 | 14 | 41 |
| $1.0 \times 10^{-6}$ | 9 | 9 | 9 | 9 | 10 | 19 | 91 |
| $1.0 \times 10^{-7}$ | 11 | 10 | 10 | 11 | 11 | 24 | 140 |
| $1.0 \times 10^{-8}$ | 12 | 12 | 12 | 12 | 13 | 28 | 189 |
| $1.0 \times 10^{-9}$ | 14 | 13 | 13 | 13 | 14 | 33 | 238 |
| $1.0 \times 10^{-10}$ | 15 | 14 | 15 | 15 | 16 | 38 | 288 |

### 3.9. Comparison of the Jacobi, Red–Black and Gauss–Seidel

We compare the performance of three relaxation methods: Jacobi, Red–Black and Gauss–Seidel. The initial condition is $\phi(x, y, 0) = 0.1\cos(\pi x)\cos(\pi y)$ on $\Omega = (0, 1) \times (0, 1)$. The parameters are $h = 1/128$, $\epsilon = 0.06$, $\Delta t = 10^{-7}$, $T = 100\Delta t$ and $tol = 1.0 \times 10^{-10}$. The SMOOTH relaxation numbers on the finest multigrid level (i.e., $\nu^1 = \nu^2$) are taken to be from 1 to 5 and those on the other multigrid levels are selected as 2 with the $128 \times 128$ mesh size. The relaxation numbers are rounded off to the nearest integer. Table 8 shows the average number of V-cycles for different relaxation numbers with the three methods. The relationship between the average numbers of V-cycles and $\nu^1 = \nu^2$ with the Jacobi, Red–Black and Gauss–Seidel method is plotted in Figure 12. The Gauss–Seidel method is observed to be the fastest. In the parallel multigrid method, the relaxation options are either Jacobi or Red–Black [33]. The Jacobi method requires approximately twice as many V-cycles as the Red–Black method does.

**Table 8.** Average numbers of V-cycles for various relaxation numbers. The SMOOTH relaxation numbers on the finest multigrid level (i.e., $\nu^1 = \nu^2$) are taken to be from 1 to 5.

| Case | Jacobi | Red–Black | Gauss-Seidal |
|---|---|---|---|
| 1 | 32 | 13 | 9 |
| 2 | 16 | 8 | 6 |
| 3 | 11 | 6 | 5 |
| 4 | 8 | 5 | 4 |
| 5 | 7 | 5 | 3 |

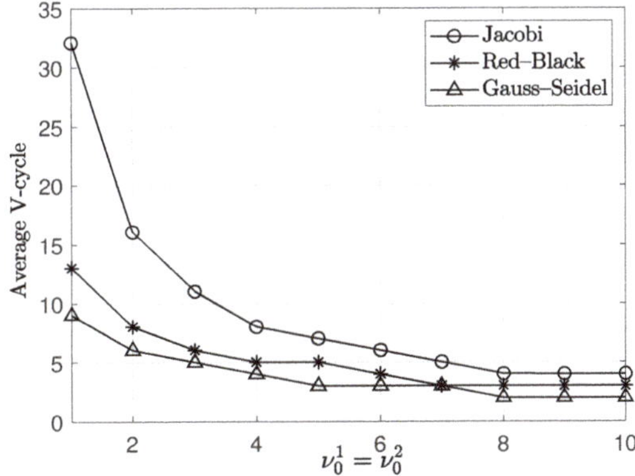


**Figure 12.** Plot of the average numbers of V-cycles versus $\nu^1 = \nu^2$ with the Jacobi (∘), Red–Black (∗) and Gauss–Seidel (△) method.

### 3.10. *Effect of* $\epsilon$

Next, we investigate the effect of $\epsilon = \epsilon_m$, which is related to the interface width. In this test, we perform a numerical simulation with the initial condition

$$\phi(x, y, 0) = \begin{cases} 1 & \text{if } 0.15 \leq x \leq 0.85 \text{ and } 0.15 \leq y \leq 0.85, \\ -1 & \text{otherwise,} \end{cases}$$

on $\Omega = (0, 1) \times (0, 1)$. We use $h = 1/128$, $\Delta t = h$, SMOOTH relaxation $= 2$, $tol = 1.0 \times 10^{-10}$ and $T = 1000\Delta t$. Figure 13 presents the evolution of the CH equation with the three values $\epsilon_4$, $\epsilon_8$ and $\epsilon_{16}$. As we have expected, the lower value of $\epsilon$ leads to a narrower interface width.

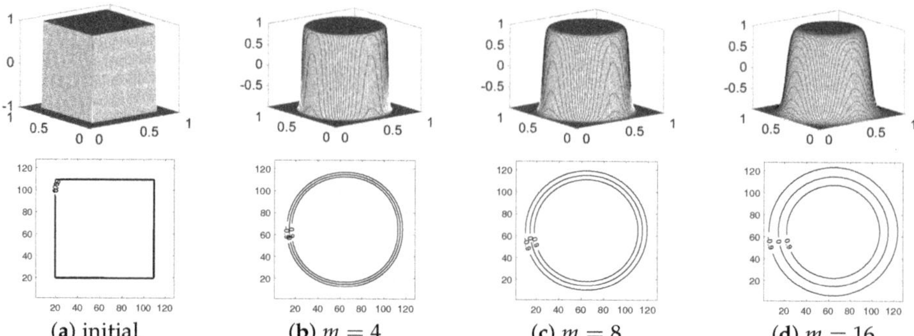

**Figure 13.** Evolution of the Cahn–Hilliard (CH) equation with different $\epsilon = \epsilon_m$: (**a**) initial condition, (**b–d**) $m = 4$, $m = 8$ and $m = 16$ at $T = 1000\Delta t$, respectively.

### 3.11. Effect of mesh size, $N_x \times N_y$

In this test, we compare the CPU times with different mesh sizes $N_x \times N_y$. The initial condition is $\phi(x, y, 0) = 0.1\cos(\pi x)\cos(\pi y)$ on $\Omega = (0, N_x/32) \times (0, N_y/32)$. The parameters are $h = 1/32$, $\Delta t = h$, $T = 100\Delta t$, $\epsilon = 0.06$, SMOOTH relaxation = 2 and $tol = 1.0 \times 10^{-10}$. Table 9 shows the CPU times and their ratios (that is, the ratio of the CPU time with the mesh size $2N_x \times 2N_y$ to the CPU time with $N_x \times N_y$). We observe that the values converge to 4.

**Table 9.** CPU times for different mesh sizes.

| Mesh Size | 32 × 32 | | 64 × 64 | | 128 × 128 | | 256 × 256 |
|---|---|---|---|---|---|---|---|
| CPU time(s) | 0.610 | | 2.812 | | 12.156 | | 50.594 |
| Ratio | | 4.610 | | 4.323 | | 4.162 | |

## 4. Conclusions

In this paper, we presented a nonlinear multigrid implementation for the CH equation in a two-dimensional space. Eyre's unconditionally gradient stable scheme was used to discretize the governing equation. The resulting discretizing equations were solved using the nonlinear multigrid method. We described the implementation of our numerical scheme in detail. We numerically showed the decrease in discrete total energy and the convergence of discrete total mass. We took a convergence test by studying the reductions in residual error on various mesh sizes in a single time step. The results of various numerical experiments were presented to demonstrate the effects of tolerance, SMOOTH relaxation, V-cycle and $\epsilon$. The provided multigrid source code will be useful to beginners who needs the numerical implementation of the nonlinear multigrid method for the CH equation.

**Author Contributions:** All authors, C.L., D.J., J.Y., and J.K., contributed equally to this work and critically reviewed the manuscript. All authors have read and agreed to the published version of the manuscript.

**Funding:** The first author (C. Lee) was supported by Basic Science Research Program through the National Research Foundation of Korea(NRF) funded by the Ministry of Education(NRF-2019R1A6A3A13094308). D. Jeong was supported by the National Research Foundation of Korea (NRF) grant funded by the Korea government (MSIP) (NRF-2017R1E1A1A03070953). J. Yang is supported by China Scholarship Council (201908260060). The corresponding author (J.S. Kim) expresses thanks for the support from the BK21 PLUS program.

**Acknowledgments:** The authors thank the editor and the reviewers for their constructive and helpful comments on the revision of this article.

**Conflicts of Interest:** The authors declare no conflicts of interest.

## Appendix A

The C code and MATLAB postprocessing code are given as follows, and the parameters are enumerated in Table A1.

**Table A1.** Parameters used for the 2D Cahn–Hilliard equation.

| Parameters | Description |
|---|---|
| nx, ny | maximum number of grid points in the x-, y-direction |
| n_level | number of multigrid level |
| c_relax | number of times being relax |
| dt | $\Delta t$ |
| xleft, yleft | minimum value on the x-, y-axis |
| xright, yright | maximum value on the x-, y-axis |
| ns | number of print out data |
| max_it | maximum number of iteration |
| max_it_mg | maximum number of multigrid iteration |
| tol_mg | tolerance for multigrid |
| h | space step size |
| h2 | $h^2$ |
| gam | $\epsilon$ |
| Cahn | $\epsilon^2$ |

The following C code is available on the following website:

http://elie.korea.ac.kr/~cfdkim/codes/

```
#include <stdio.h>
#include <math.h>
#include <stdlib.h>
#include <malloc.h>
#include <time.h>
#define gnx 32
#define gny 32
#define PI 4.0*atan(1.0)
#define iloop for(i=1;i<=gnx;i++)
#define jloop for(j=1;j<=gny;j++)
#define ijloop iloop jloop
#define iloopt for(i=1;i<=nxt;i++)
#define jloopt for(j=1;j<=nyt;j++)
#define ijloopt iloopt jloopt
int nx,ny,n_level,c_relax;
double **ct,**sc,**smu,**sor,h,h2,dt,xleft,xright,yleft,yright,gam,Cahn,**mu,**mi;
double **dmatrix(long nrl,long nrh,long ncl,long nch){
double **m;
long i,nrow=nrh-nrl+2,ncol=nch-ncl+2;
m=(double **) malloc((nrow)*sizeof(double*)); m+=1;m-=nrl;
m[nrl]=(double *) malloc((nrow*ncol)*sizeof(double)); m[nrl]+=1; m[nrl]-=ncl;
for (i=nrl+1; i<=nrh; i++) m[i]=m[i-1]+ncol;
return m;
}
void free_dmatrix(double **m,long nrl,long nrh,long ncl,long nch){
free(m[nrl]+ncl-1); free(m+nrl-1);
}
void zero_matrix(double **a,int xl,int xr,int yl,int yr){
int i,j;
```

```
for (i=xl; i<=xr; i++){ for (j=yl; j<=yr; j++){ a[i][j]=0.0; }}
}
```

```
void mat_add2(double **a,double **b,double **c,double **a2,
double **b2,double **c2,int xl,int xr,int yl,int yr){
int i,j;
for (i=xl; i<=xr; i++)
for (j=yl; j<=yr; j++){ a[i][j]=b[i][j]+c[i][j]; a2[i][j]=b2[i][j]+c2[i][j]; }
}
void mat_sub2(double **a,double **b,double **c,double **a2,
double **b2,double **c2,int nrl,int nrh,int ncl,int nch){
int i,j;
for (i=nrl;i<=nrh;i++)
for (j=ncl; j<=nch; j++){ a[i][j]=b[i][j]-c[i][j];a2[i][j]=b2[i][j]-c2[i][j]; }
}
void mat_copy(double **a,double **b,int xl,int xr,int yl,int yr){
int i,j;
for (i=xl; i<=xr; i++){ for (j=yl; j<=yr; j++){ a[i][j]=b[i][j]; }}
}
void mat_copy2(double **a,double **b,double **a2,double **b2,int xl,int xr,int yl,int yr){
int i,j;
for (i=xl; i<=xr; i++)
for (j=yl; j<=yr; j++){a[i][j]=b[i][j]; a2[i][j]=b2[i][j];}
}
void print_mat(FILE *fptr,double **a,int nrl,int nrh,int ncl,int nch){
int i,j;
for(i=nrl; i<=nrh; i++){ for(j=ncl; j<=nch; j++)
fprintf(fptr," %16.15f",a[i][j]); fprintf(fptr,"\n"); }
}
void print_data(double **phi) {
FILE *fphi;
fphi=fopen("phi.m","a"); print_mat(fphi,phi,1,nx,1,ny); fclose(fphi);
}
void laplace(double **a,double **lap_a,int nxt,int nyt){
int i,j;
double ht2,dadx_L,dadx_R,dady_B,dady_T;
ht2=pow((xright-xleft)/(double) nxt,2);
ijloopt {
if (i>1)   dadx_L=a[i][j]-a[i-1][j];
else       dadx_L=0.0;
if (i<nxt) dadx_R=a[i+1][j]-a[i][j];
else       dadx_R=0.0;
if (j>1)   dady_B=a[i][j]-a[i][j-1];
else       dady_B=0.0;
if (j<nyt) dady_T=a[i][j+1]-a[i][j];
else       dady_T=0.0;
lap_a[i][j]=(dadx_R-dadx_L+dady_T-dady_B)/ht2;}
}
void source(double **c_old,double **src_c,double **src_mu){
int i,j;
laplace(c_old,ct,nx,ny);
```

```
ijloop{src_c[i][j]=c_old[i][j]/dt-ct[i][j]; src_mu[i][j]=0.0;}
}
double df(double c){return pow(c,3);}
```

```
double d2f(double c){return 3.0*c*c;}
void relax(double **c_new,double **mu_new,double **su,double **sw,int ilevel,
int nxt, int nyt){
int i,j,iter;
double ht2,x_fac,y_fac,a[4],f[2],det;
ht2=pow((xright-xleft)/(double) nxt,2);
for (iter=1; iter<=c_relax; iter++){
ijloopt {
if (i>1 && i<nxt) x_fac=2.0;
else              x_fac=1.0;
if (j>1 && j<nyt) y_fac=2.0;
else              y_fac=1.0;
a[0]=1.0/dt; a[1]=(x_fac+y_fac)/ht2;
a[2]=-(x_fac+y_fac)*Cahn/ht2-d2f(c_new[i][j]); a[3]=1.0;
f[0]=su[i][j]; f[1]=sw[i][j]+df(c_new[i][j])-d2f(c_new[i][j])*c_new[i][j];
if (i>1) { f[0]+=mu_new[i-1][j]/ht2; f[1]-=Cahn*c_new[i-1][j]/ht2; }
if (i<nxt) { f[0]+=mu_new[i+1][j]/ht2; f[1]-=Cahn*c_new[i+1][j]/ht2; }
if (j>1) { f[0]+=mu_new[i][j-1]/ht2; f[1]-=Cahn*c_new[i][j-1]/ht2; }
if (j<nyt) { f[0]+=mu_new[i][j+1]/ht2; f[1]-=Cahn*c_new[i][j+1]/ht2; }
det=a[0]*a[3]-a[1]*a[2];
c_new[i][j]=(a[3]*f[0]-a[1]*f[1])/det;
mu_new[i][j]=(-a[2]*f[0]+a[0]*f[1])/det; }}
}
void restrictCH(double **uf,double **uc,double **vf,double **vc,int nxc,int nyc) {
int i,j;
for (i=1; i<=nxc; i++)
for (j=1; j<=nyc; j++){
uc[i][j]=0.25*(uf[2*i][2*j]+uf[2*i-1][2*j]+uf[2*i][2*j-1]+uf[2*i-1][2*j-1]);
vc[i][j]=0.25*(vf[2*i][2*j]+vf[2*i-1][2*j]+vf[2*i][2*j-1]+vf[2*i-1][2*j-1]);}
}
void nonL(double **ru,double **rw,double **c_new,double **mu_new,int nxt,int nyt) {
int i,j;
double **lap_mu,**lap_c;
lap_mu=dmatrix(1,nxt,1,nyt); lap_c=dmatrix(1,nxt,1,nyt);
laplace(c_new,lap_c,nxt,nyt); laplace(mu_new,lap_mu,nxt,nyt);
ijloopt{ ru[i][j]=c_new[i][j]/dt-lap_mu[i][j];
rw[i][j]=mu_new[i][j]-df(c_new[i][j])+Cahn*lap_c[i][j]; }
free_dmatrix(lap_mu,1,nxt,1,nyt); free_dmatrix(lap_c,1,nxt,1,nyt);
}
void defect(double **duc,double **dwc,double **uf_new,double **wf_new,double **suf,
double **swf,int nxf,int nyf,double **uc_new,double **wc_new,int nxc,int nyc) {
double **ruf,**rwf,**rruf,**rrwf,**ruc,**rwc;
ruc=dmatrix(1,nxc,1,nyc);rwc=dmatrix(1,nxc,1,nyc);ruf=dmatrix(1,nxf,1,nyf);
rwf=dmatrix(1,nxf,1,nyf);rruf=dmatrix(1,nxc,1,nyc);rrwf=dmatrix(1,nxc,1,nyc);
nonL(ruc,rwc,uc_new,wc_new,nxc,nyc);nonL(ruf,rwf,uf_new,wf_new,nxf,nyf);
mat_sub2(ruf,suf,ruf,rwf,swf,rwf,1,nxf,1,nyf);
restrictCH(ruf,rruf,rwf,rrwf,nxc,nyc);
```

```
mat_add2(duc,ruc,rruf,dwc,rwc,rrwf,1,nxc,1,nyc);
free_dmatrix(ruc,1,nxc,1,nyc); free_dmatrix(rwc,1,nxc,1,nyc);
free_dmatrix(ruf,1,nxf,1,nyf); free_dmatrix(rwf,1,nxf,1,nyf);
free_dmatrix(rruf,1,nxc,1,nyc); free_dmatrix(rrwf,1,nxc,1,nyc);
}
```

```
void prolong_ch(double **uc,double **uf,double **vc,double **vf, int nxc,int nyc){
int i,j;
for (i=1; i<=nxc; i++)
for (j=1; j<=nyc; j++){
uf[2*i][2*j]=uf[2*i-1][2*j]=uf[2*i][2*j-1]=uf[2*i-1][2*j-1]=uc[i][j];
vf[2*i][2*j]=vf[2*i-1][2*j]=vf[2*i][2*j-1]=vf[2*i-1][2*j-1]=vc[i][j];}
}
void vcycle(double **uf_new,double **wf_new,double **su,double **sw,int nxf,int nyf,
int ilevel) {
relax(uf_new,wf_new,su,sw,ilevel,nxf,nyf);
if (ilevel<n_level) {
int nxc,nyc;
double **duc,**dwc,**uc_new,**wc_new,**uc_def,**wc_def,**uf_def,**wf_def;
nxc=nxf/2; nyc=nyf/2;
duc=dmatrix(1,nxc,1,nyc); dwc=dmatrix(1,nxc,1,nyc);
uc_new=dmatrix(1,nxc,1,nyc); wc_new=dmatrix(1,nxc,1,nyc);
uf_def=dmatrix(1,nxf,1,nyf); wf_def=dmatrix(1,nxf,1,nyf);
uc_def=dmatrix(1,nxc,1,nyc); wc_def=dmatrix(1,nxc,1,nyc);
restrictCH(uf_new,uc_new,wf_new,wc_new,nxc,nyc);
defect(duc,dwc,uf_new,wf_new,su,sw,nxf,nyf,uc_new,wc_new,nxc,nyc);
mat_copy2(uc_def,uc_new,wc_def,wc_new,1,nxc,1,nyc);
vcycle(uc_def,wc_def,duc,dwc,nxc,nyc,ilevel+1);
mat_sub2(uc_def,uc_def,uc_new,wc_def,wc_def,wc_new,1,nxc,1,nyc);
prolong_ch(uc_def,uf_def,wc_def,wf_def,nxc,nyc);
mat_add2(uf_new,uf_new,uf_def,wf_new,wf_new,wf_def,1,nxf,1,nyf);
relax(uf_new,wf_new,su,sw,ilevel,nxf,nyf);
free_dmatrix(duc,1,nxc,1,nyc); free_dmatrix(dwc,1,nxc,1,nyc);
free_dmatrix(uc_new,1,nxc,1,nyc); free_dmatrix(wc_new,1,nxc,1,nyc);
free_dmatrix(uf_def,1,nxf,1,nyf); free_dmatrix(wf_def,1,nxf,1,nyf);
free_dmatrix(uc_def,1,nxc,1,nyc); free_dmatrix(wc_def,1,nxc,1,nyc); }
}
double error2(double **c_old,double **c_new,double **mu,int nxt,int nyt){
int i,j;
double **rr,res2,x=0.0;
rr=dmatrix(1,nxt,1,nyt);
ijloopt { rr[i][j]=mu[i][j]-c_old[i][j]; }
laplace(rr,sor,nx,ny);
ijloopt { rr[i][j]=sor[i][j]-(c_new[i][j]-c_old[i][j])/dt; }
ijloopt { x=(rr[i][j])*(rr[i][j])+x; }
res2=sqrt(x/(nx*ny));
free_dmatrix(rr,1,nxt,1,nyt);
return res2;
}
void initialization(double **phi){
int i,j;
```

```
double x,y;
ijloop {x=(i-0.5)*h; y=(j-0.5)*h; phi[i][j]=cos(PI*x)*cos(PI*y);}
}
void cahn(double **c_old,double **c_new){
FILE *fphi2;
int i,j,max_it_CH=10000,it_mg2=1;
```

```
double tol=1.0e-10, resid2=1.0;
source(c_old,sc,smu);
while (it_mg2<=max_it_CH && resid2>tol) {
it_mg2++; vcycle(c_new,mu,sc,smu,nx,ny,1);
resid2=error2(c_old,c_new,mu,nx,ny);
printf("error2 %16.15f %d \n",resid2,it_mg2-1);
fphi2=fopen("phi2.m","a");
fprintf(fphi2,"%16.15f %d \n",resid2,it_mg2-1); fclose(fphi2);}
}
int main(){
int it=1,max_it,ns,count=1,it_mg=1;
double **oc,**nc,resid2=1.0;
FILE *fphi,*fphi2;
c_relax=2; nx=gnx; ny=gny; n_level=(int)(log(nx)/log(2.0)+0.1);
xleft=0.0; xright=1.0; yleft=0.0; yright=1.0; max_it=100; ns=10; dt=0.01;
h=xright/(double)nx; h2=pow(h,2); gam=0.06; Cahn=pow(gam,2);
printf("nx=%d,ny=%d\n",nx,ny); printf("dt=%f\n",dt);
printf("max_it=%d\n",max_it); printf("ns=%d\n",ns); printf("n_level=%d\n\n",n_level);
oc=dmatrix(0,nx+1,0,ny+1); nc=dmatrix(0,nx+1,0,ny+1); mu=dmatrix(1,nx,1,ny);
sor=dmatrix(1,nx,1,ny); ct=dmatrix(1,nx,1,ny); sc=dmatrix(1,nx,1,ny);
mi=dmatrix(1,nx,1,ny); smu=dmatrix(1,nx,1,ny); zero_matrix(mu,1,nx,1,ny);
initialization(oc); mat_copy(nc,oc,1,nx,1,ny);
fphi=fopen("phi.m","w"); fclose(fphi); print_data(oc);
for (it=1; it<=max_it; it++) {
cahn(oc,nc); mat_copy(oc,nc,1,nx,1,ny);
if (it % ns==0) {count++; print_data(oc); printf("print out counts %d \n",count);}
printf(" %d \n",it);}
return 0;
}
```

The following MATLAB code produces the results shown in Figure 5. The code can also be downloaded from

http://elie.korea.ac.kr/~cfdkim/codes/

```
clear; clc; close all;
ss=sprintf('./phi.m'); phi=load(ss); nx=32; ny=32; n=size(phi,1)/nx;
x=linspace(0,1,nx); y=linspace(0,1,ny); [xx,yy]=meshgrid(x,y);
for i=1:n
pp=phi((i-1)*nx+1:i*nx,:);
figure(i); mesh(xx,yy,pp'); axis([0 1 0 1 -1 1]); view(-38,42);
end
```

## References


1. Trottenberg, U.; Schüller, A.; Oosterlee, C.W. *Multigrid Methods*; Academic Press: Cambridge, MA, USA, 2000.
2. Cahn, J.W.; Hilliard, J.E. Free energy of a nonuniform system. I. Interfacial free energy. *J. Chem. Phys.* **1958**, *28*, 258–267. [CrossRef]
3. Copetti, M.I.M.; Elliott, C.M. Kinetics of phase decomposition processes: Numerical solutions to Cahn–Hilliard equation. *Mater. Sci. Technol.* **1990**, *6*, 273–284. [CrossRef]
4. Honjo, M.; Saito, Y. Numerical simulation of phase separation in Fe-Cr binary and Fe-Cr-Mo ternary alloys with use of the Cahn–Hilliard equation. *ISIJ Int.* **2000**, *40*, 914–919. [CrossRef]
5. Bertozzi, A.L.; Esedoglu, S.; Gillette, A. Inpainting of binary images using the Cahn–Hilliard equation. *IEEE Trans. Image Process.* **2007**, *16*, 285–291. [CrossRef]
6. Bosch, J.; Kay, D.; Stoll, M.; Wathen, A.J. Fast solvers for Cahn–Hilliard inpainting. *SIAM J. Imaging Sci.* **2014**, *7*, 67–97. [CrossRef]
7. Choksi, R.; Peletier, M.A.; Williams, J.F. On the phase diagram for microphase separation of diblock copolymers: An approach via a nonlocal Cahn–Hilliard functional. *SIAM J. Appl. Math.* **2009**, *69*, 1712–1738. [CrossRef]
8. Tang, P.; Qiu, F.; Zhang, H.; Yang, Y. Phase separation patterns for diblock copolymers on spherical surfaces: A finite volume method. *Phys. Rev. E* **2005**, *72*, 016710. [CrossRef] [PubMed]
9. Hu, S.Y.; Chen, L.Q. A phase-field model for evolving microstructures with strong elastic inhomogeneity. *Acta Mater.* **2001**, *49*, 1879–1890. [CrossRef]
10. Yu, P.; Hu, S.Y.; Chen, L.Q.; Du, Q. An iterative-perturbation scheme for treating inhomogeneous elasticity in phase-field models. *J. Comput. Phys.* **2005**, *208*, 34–50. [CrossRef]
11. Gurtin, M.E.; Polignone, D.; Vinals, J. Two-phase binary fluids and immiscible fluids described by an order parameter. *Math. Models Methods Appl. Sci.* **1996**, *6*, 815–831. [CrossRef]
12. Lee, T. Effects of incompressibility on the elimination of parasitic currents in the lattice Boltzmann equation method for binary fluids. *Comput. Math. Appl.* **2009**, *58*, 987–994. [CrossRef]
13. Jeong, D.; Kim, J. Phase-field model and its splitting numerical scheme for tissue growth. *Appl. Numer. Math.* **2017**, *117*, 22–35. [CrossRef]
14. Kim, J.; Lee, S.; Choi, Y.; Lee, S.M.; Jeong, D. Basic Principles and Practical Applications of the Cahn–Hilliard Equation. *Math. Probl. Eng.* **2016**, *2016*, 9532608. [CrossRef]
15. Colli, P.; Gilardi, G.; Sprekels, J. A distributed control problem for a fractional tumor growth model. *Mathematics* **2019**, *7*, 792. [CrossRef]
16. Myśliński, A.; Wroblewski, M. Structural optimization of contact problems using Cahn–Hilliard model. *Comput. Struct.* **2017**, *180*, 52–59. [CrossRef]
17. Eyre, D.J. Unconditionally gradient stable time marching the Cahn–Hilliard equation. *MRS Proc.* **1998**, *529*, 39. [CrossRef]
18. Yang, S.; Lee, H.; Kim, J. A phase-field approach for minimizing the area of triply periodic surfaces with volume constraint. *Comput. Phys. Commun.* **2010**, *181*, 1037–1046. [CrossRef]
19. Kim, J. A numerical method for the Cahn–Hilliard equation with a variable mobility. *Commun. Nonlinear Sci. Numer. Simul.* **2007**, *12*, 1560–1571. [CrossRef]
20. Shin, J.; Jeong, D.; Kim, J. A conservative numerical method for the Cahn–Hilliard equation in complex domains. *J. Comput. Phys.* **2011**, *230*, 7441–7455. [CrossRef]
21. Jeong, D.; Kim, J. A practical numerical scheme for the ternary Cahn–Hilliard system with a logarithmic free energy. *Phys. A* **2016**, *442*, 510–522. [CrossRef]
22. Lee, H.; Kim, J. A second-order accurate non-linear difference scheme for the N-component Cahn–Hilliard system. *Phys. A* **2008**, *387*, 4787–4799. [CrossRef]
23. Lee, H.G.; Shin, J.; Lee, J.Y. A High-Order Convex Splitting Method for a Non-Additive Cahn–Hilliard Energy Functional. *Mathematics* **2019**, *7*, 1242. [CrossRef]
24. Shin, J.; Choi, Y.; Kim, J. An unconditionally stable numerical method for the viscous Cahn–Hilliard equation. *Discret. Contin. Dyn. Syst. Ser. B* **2014**, *19*, 1737–1747. [CrossRef]
25. Kim, J. Phase-field models for multi-component fluid flows. *Commun. Comput. Phys.* **2012**, *12*, 613–661. [CrossRef]

26. Kim, J.; Bae, H.O. An unconditionally gradient stable adaptive mesh refinement for the Cahn–Hilliard equation. *J. Korean Phys. Soc.* **2008**, *53*, 672–679. [CrossRef]
27. Wise, S.; Kim, J.; Lowengrub, J. Solving the regularized, strongly anisotropic Cahn–Hilliard equation by an adaptive nonlinear multigrid method. *J. Comput. Phys.* **2007**, *226*, 414–446. [CrossRef]
28. Li, Y.; Jeong, D.; Shin, J.; Kim, J. A conservative numerical method for the Cahn–Hilliard equation with Dirichlet boundary conditions in complex domains. *Comput. Math. Appl.* **2013**, *65*, 102–115. [CrossRef]
29. Lee, H.G.; Kim, J. Accurate contact angle boundary conditions for the Cahn–Hilliard equations. *Comput. Fluids* **2011**, *44*, 178–186. [CrossRef]
30. Shin, J.; Kim, S.; Lee, D.; Kim, J. A parallel multigrid method of the Cahn–Hilliard equation. *Comput. Mater. Sci.* **2013**, *71*, 89–96. [CrossRef]
31. Lee, C.; Jeong, D.; Shin, J.; Li, Y.; Kim, J. A fourth-order spatial accurate and practically stable compact scheme for the Cahn–Hilliard equation. *Phys. A* **2014**, *409*, 17–28. [CrossRef]
32. Choi, J.W.; Lee, H.G.; Jeong, D.; Kim, J. An unconditionally gradient stable numerical method for solving the Allen–Cahn equation. *Phys. A* **2009**, *388*, 1791–1803. [CrossRef]
33. Baker, A.H.; Falgout, R.D.; Kolev, T.V.; Yang, U.M. Scaling hypre's multigrid solvers to 100,000 cores. In *High-Performance Scientific Computing*; Springer: London, UK, 2012; pp. 261–279.

MDPI
St. Alban-Anlage 66
4052 Basel
Switzerland
Tel. +41 61 683 77 34
Fax +41 61 302 89 18
www.mdpi.com

*Mathematics* Editorial Office
E-mail: mathematics@mdpi.com
www.mdpi.com/journal/mathematics

www.ingramcontent.com/pod-product-compliance
Lightning Source LLC
LaVergne TN
LVHW070630170726
843515LV00004B/221
*9783039287260*